An Introduction to Regional Geography

PAUL CLAVAL

Translated by Ian Thompson

First published in French as *Initiation à la Géographie Régionale* by Éditions Nathan in 1993.

First published in English 1998

This English edition has been published with the support of the French Ministry of Culture.

2 4 6 8 10 9 7 5 3 1

Blackwell Publishers Ltd
108 Cowley Road
Oxford OX4 1JF
UK

Blackwell Publishers Inc.
350 Main Street
Malden, Massachusetts 02148
USA

British Library Cataloguing in Publication Data
A CIP catalogue record for this book is available from the British Library.

Library of Congress Cataloging-in-Publication Data
Claval, Paul.
[Initiation à la géographie régionale. English]
An introduction to regional geography / Paul Claval : translated by Ian Thompson.
p. cm.
Includes bibliographical references and index.
ISBN 1-55786-732-1. — ISBN 1-55786-733-X (pbk.)
1. Geography. I. Title.
G116.C48413 1998
910—dc21 97-32760
CIP

Typeset in 11 on 13 pt Plantin
by Graphicraft Typesetters Ltd., Hong Kong
Printed in Great Britain by T. J. International Ltd., Padstow, Cornwall

This book is printed on acid-free paper

An Introduction to Regional Geography

Contents

Figures

Translator's Preface

It is highly appropriate that an overview of regional geography should be presented by a distinguished French geographer, since France occupied a distinctive place in the evolution of the regional concept. From the early years of the present century until the 1950s, what may be termed the 'classical' school of French human and regional geography, or more specifically, the Vidalian school after its leading exponent, Paul Vidal de la Blache, enjoyed an international influence in both teaching and research (Crone, 1951; Harrison-Church, 1957; Sister Mary Annette, 1965; Wrigley, 1965; Freeman, 1967; Ribeiro, 1968). In terms of the regional concept, the Vidalian approach held the view that each fragment of the earth's space contained its own internal logic as far as the physical environment and human response were concerned (Vidal de la Blache, 1910). The model in France for this possibilist and holistic approach was the *pays*, a small-scale unit corresponding to a distinctive landscape assemblage based on the physical environment and the human adaptation to its opportunities and constraints. On the one hand, the cultural landscape was the manifestation of these linkages, but in addition, the *pays* involved a strong sense of identification on the part of its inhabitants. *Pays* were thus defined by a way of life as well as by the sometimes arbitrary divisions based on geological boundaries or historical affinity. The flowering of this approach was expressed in a whole suite of regional monographs, produced by contemporaries and disciples of Vidal, characterized by scholarly depth, elegance of style and an historical approach which elucidated the formation of cultural landscapes. In a country where great tracts of land remained largely untransformed by industrialization and with relatively discrete urban influences, it was possible to propose such regions as having a high degree of homogeneity. The Vidalian approach to landscape description and analysis thus came to represent a kind of orthodoxy, perpetuated by the succeeding generation of scholars in what became a distinctive French national school of human and regional geography. Although Vidal foreshadowed a more functional approach based on the polarizing effects of towns, such was the weight of the tradition, as expressed in university and high school curricula, basic texts and research monographs, that the classical geographical regional approach persisted until the trauma of the Second World War and the rapid post-war transformation of the French economy when society called into question the relevance of a relatively inert methodology of regionalization to a dynamic French space economy. It has been argued that the traditional geographical regional approach proceeded by replication rather than by a close examination of the conceptual framework, or increased rigour in methodology. In particular, the idiographic method did not lend itself to generalizations, principles, laws or theories (Juillard, 1962). The scholarly effort of establishing the distinctive 'personality' of areas was ill-adapted to accommodate a post-war France of new industries, demographic upsurge and a volatile society, which inevitably effected fundamental spatial changes. Thus, the 1960s heralded an active debate in France on the purpose and methods of regional geography and, more widely, on the substance of human geography (Claval, 1964, 1975; Claval and Thompson, 1975). It was to be a debate which was prolonged, polemical and not without conflict. It must be stressed that this debate was not a total renunciation of the classical traditional approach of the pre-war decades, which was still held in esteem as a golden age, when French geography enjoyed international pre-eminence. Rather, the task was to map out a new agenda and new methodologies to suit a France with a modernized economy and an increasingly urbanized society, which no longer reposed on arbitrarily defined regions of supposed homogeneity. It was also apparent, especially to a new generation of geographers, that major philosophical changes and innovative methodologies were being introduced abroad and especially in the Anglo-Saxon world, which implied that the persistence of 'National' schools was outmoded (Claval and Thompson, 1975). The spirit of change thus had to come to terms with two imperatives. Internally the rapid transformation of France thrust upon geographers new issues, new subject matter and new research priorities. Externally, French geographers had to accommodate ideas emanating from abroad and in particular the abandonment of the geographical region in its traditional form in favour of more varied and flexible approaches to spatial organization.

Several strands may be identified in the reappraisal of French regional geography. The creation of official administration and planning regions after 1955 by the government stimulated debate and reaction focused on the arbitrary nature of the boundaries adopted (Labasse, 1960). This usurping of the word 'region' undoubtedly was a challenge to geographical orthodoxy, the more so when it became increasingly clear that the new regions were to be a basis for action and decision-making rather than simply for administrative convenience. Similarly, the projection of Perrouxian economic theory into spatial dimensions by Boudeville produced the functional concept of polarized space, which seemed more advanced than traditional regional theory in that it involved an operational approach that could be modelled statistically and potentially had a predictive value (Boudeville, 1961). The diffusion of quantitative methods represented a further challenge, especially in view of its Anglo-Saxon origins. Initial resistance was all the more contradictory in that the demographic and socio-economic databases provided by the *Institut National de la Statistique et des Études Economiques* (INSEE) are amongst the most detailed and comprehensive in Europe. Moreover, the data produced by INSEE increasingly appeared in aggregated forms that corresponded to functional spatial structures, for example, urban-industrial units (multi-communal agglomerations) and zones of urban influence in terms of labour catchments. In the traditional regional approach, statistical information had been largely involved to underpin description, but by the 1960s, statistical methods were being used for analysis and for the process of regionalization and classification, especially in terms of urban systems (J. Hautreux and M. Rochefort, 1965). Thus, the use of statistics progressed from being a support for idiographic writing to being an investigative tool and to a more nomothetic approach linking fact to theory. The quest for theory-building opened the door to a more direct challenge to orthodoxy in that it was explicitly political and philosophical in nature and, to a degree, more structural than spatial. The impact of Marxist theory in the social sciences generally, inevitably found an echo in geographical thought and the appearance of a new journal, *Hérodote*, in 1976, supplied an outlet for radical and critical geography in much the same manner as its antecedent, *Antipode*, had provided in the Anglo-Saxon domain in 1969. In effect, from the 1970s onwards it became increasingly difficult to identify a French national school of geography but rather a progressive integration of French geography into the international currents of geographical thought. This process was facilitated by the translation into French of a number of landmark texts of British and American origin (for example, P. Haggett, 1973) and there is little doubt that Claval was a major catalyst in this sea-change in epistemology (Claval, 1977). This new ferment was epitomized in 1972 by the appearance of a new journal, *L'Espace Géographique*, explicitly conceived as being a platform for critical discussion, international in spirit and innovative in format.

Whilst at this macro scale an intense period of debate had virtually consigned traditional regional geography to the status of a nostalgic memory, at a domestic scale a further debate was to re-focus the regional concept at a more practical level, again not without certain tensions. The classical school of French regional geography had been essentially academic with only faint hints that the geographer's skills might have a directly applied value. Nevertheless, a proud tradition, which still persists, in French universities, was a devotion to research within their regional hinterlands. With the government's adoption of formal regional planning together with the creation of the *Délégation à l'Aménagement du Territoire*, a high-level government agency devoted to planning and coordinating major development policies and projects, it was natural for geographers to engage with the development problems of their local regions and thus with an applied approach. This endowed certain benefits: the introduction of new subject matter eventually reflected in new texts, the inclusion of geographers in multi-disciplinary teams, new sources of research funding, new career opportunities for geography graduates and an opportunity to demonstrate the relevance of the discipline to society at large. While opening a new avenue for the expression of regional expertise, the applied approach was not adopted without controversy (McDonald, 1964). For some, the time restriction imposed by contract research encouraged geographers to reach hasty conclusions that more patient research would contradict and that would derogate young researchers to the status of technicians (George, 1961). From this point of view, the geographer should act as analyst and commentator rather than practitioner (George, 1964). The opposing view, advanced by Phlipponneau (Phlipponneau, 1960), considered that geographers should be actively and politically engaged in issues of regional development. Others considered this conflict to be sterile since both points of view produced geographical analyses which were applicable at the regional scale (Juillard, 1965). At its best, applied regional study retained key features of the classical approach – spatial awareness, an integrative approach and profound knowledge of the region concerned – but added new dimensions. Chief of these were a problem-based rationale, greater interaction with practitioners from other disciplines and an enhanced vision of space as consisting of overlapping, inter-digitating and frequently conflicting socio-economic systems.

The various strands of debate which have been identified above represent just a selection of the currents that have influenced regional geography in France. By the present decade, it has emerged as being no longer monolithic, or as some would consider, narcissistic. The notion of a specifically French School has been replaced by an openness to developments in the discipline at large, to a breaking down of disciplinary boundaries and to increased rigour in both philosophy and methodology. In this process of modernization, Professor Claval has played a pivotal role as commentator, critic and catalyst. His erudition, across a range of disciplines and literature in several languages, has exposed his readership to developments in geographical thought beyond the frontiers of France. In the space of thirty years he has published a prolific stream of over twenty major texts in Economic, Social, Historical and Cultural Geography and on the history of geographical thought (Claval, 1996). In personal correspondence with this translator over twenty years ago, Claval regretted the fact that for many of his colleagues he was seen simply as a transmitter of Anglo-Saxon paradigms, whereas he saw his mission as preserving what is best in the French tradition but to cross-fertilize it with ideas from abroad. His quest was not to devise original theory but to construct philosophical and practical frameworks, often at a global scale, within which knowledge can be organized more effectively. Given that the region, however elusive in theoretical terms and motley in application, has been one of the geographer's principal tools by which to organize and analyse phenomena in space, it was inevitable that he should produce an overview of this theme. The task was all the more urgent in that the decline of Vidalian approach and the intellectual turbulence in French geography post-1950 has led to an increasing lack of clarity in the regional concept in both theory and practice.

This book is essentially a work of reflection and re-statement seeking to consolidate the framework of regional study from its early origins to its modern form. It does not seek to explore the new research frontiers that currently challenge spatial constructs but that have still to crystallize into a new corpus of regional theory. For example, the re-examination of colonial geographies, issues of gendered space, geographies of exclusion, the contestation of space and domination, the formation of localities, are touched on but not developed. It is essentially a work of introduction, as the title states. The text is an integral translation of Professor Claval's *Initiation à la Géographie Régionale*, published by Nathan, in Paris, in 1993. For this English edition, Professor Claval has substantially extended chapter 1 to take into account more recent developments in geographical thought and to incorporate reading suggestions in English. The translation adheres as closely as possible to the original text but explanatory footnotes have been added to elucidate allusions that might otherwise have required further reference work by the reader. Certain French terms have no English equivalent or can only be translated by phrases. Such instances have also been explained in footnotes. All literature cited in the text has been included but some of the further reading in French has been excluded as being difficult to access or involving a greater depth of reading than will be required by most English-speaking users. Conversely, reference to the most recent literature in English has been appended, where appropriate, to allow the reader to project forwards to current research issues.

To translate a work of such scope and sophistication has been a very difficult task, particularly in terms of preserving as closely as possible the style of the original without sacrificing clarity. Nevertheless, it is hoped that this translation will contribute to the study of the evolution of the region; a concept so often considered to have exhausted its utility, but which, as Professor Claval argues, still continues to have validity, albeit in new guises.

Ian Thompson

Further Reading

Boudeville, J., *Les espaces économiques*, Paris: PUF, 1961.

Claval, P., *Essai sur l'évolution de la géographie humaine*, Paris: Les Belles Letters, 1964.

——, Contemporary human geography in France, *Progress in Geography*, vol. 7, 1975, 255–92.

——, *La nouvelle géographie*, Paris: PUF, 1977.

—— and Thompson, I. B., Trends in human geography in Britain and France, *Geographical Journal*, vol. 141, 1975, 345–54.

Claval, P., *La géographie comme genre de vie*, Paris, L'Marmattan, 1996.

Crone, G., *Modern Geographers*, London: Royal Geographical Society, 1951.

Freeman, T., *The Geographer's Craft*, Manchester: Manchester University Press, 1967.

George, P., Existe-t-il une géographie appliquée?, *Annales de Géographie*, vol. 70, 1961, 337–46.

George, P., *La géographie active*, Paris: PUF, 1964.

Harrison-Church, R., The French school of Geography. In Taylor, G., *Geography in the Twentieth Century*, Methuen, 1957.

Haggett, P., *L'analyse spatiale en géographie humaine*, Paris: Armand Colin, 1973.

Hautreux, J. and Rochefort, M., Physionomie générale de l'armature urbaine française, *Annales de Géographie*, vol. 74, 1965, 660–77.

Juillard, E., La région: essai de définition, *Annales de Géographie*, vol. 71, 1962, 483–99.

——, La région, cadre de la géographie active, *Annales de Géographie*, vol. 71, 1965, 736–8.

Labasse, J., La portée géographique des programmes d'action régionale français, *Annales de Géographie*, vol. 69, 371–93.

McDonald, J., Current controversy in French geography, *Professional Geography*, vol. 16, no. 6, 1964, 20–3.

Phlipponneau, M., *Géographie et action. Introduction à la Géographie appliqué*, Paris: Armand Colin, 1960.

Ribeiro, O., En relisant Vidal de la Blache, *Annales de Géographie*, vol. 77, 1968, 641–62.

Sister Mary Annette, The changing French region, *Professional Geography*, vol. 17, 1965, 1–5.

Vidal de la Blache, P., Régions françaises, *Revue de Paris*, vol. 17, 1910, 821–49.

Wrigley, A., Changes in the philosophy of geography. In Chorley, R. and Haggett, P. (eds), *Frontiers in Geographical Teaching*, London: Methuen, 1965.

Introduction

Early in the morning the plane takes off from Newark airport. The moment it leaves the ground, the view opens out obliquely over suburban houses, the impeccable green of their lawns, the flame-red foliage of their maples – we are in an Indian summer – the blue flashes of swimming pools, numerous in this comfortably-off district, and the cars parked on the wide blacktop in front of the houses. The plane climbs quickly, and already we have a vertical view of the precise geometry of streets and housing lots, the buildings dwindling so fast that they look like dolls' houses. Farther off, Newark Bay sparkles against the sunlight and, beyond a dark outline of suburbs, we can make out the Hudson. The Manhattan skyline, with the two tall sectors of the Midtown and Downtown dominated by the bold towers of the World Trade Center, takes shape against a background set ablaze by the rising sun. Slowly the plane turns west as it gains altitude. The whole northern suburbs of New York are fantastically compressed and foreshortened, but already we are leaving them behind. The layout of isolated dwellings and scattered housing lots is still dense in this suburban New Jersey, set in grasslands and woods. We are now at a height of about 4 or 5,000 metres, and ahead of us the landscape is cut by the first of a succession of rigid barriers. When we are directly above these Appalachian ridges, we are struck by the general alignment of their crests; they are barely interrupted, except here and there by a river gap; some of the heights form loops; the contrast between the wooded hills and the geometric patchwork of the valley floors is striking.

It takes but two or three minutes to cross the Appalachians. The plane is now at its cruising height of over 10,000 metres, and is flying over an immense, almost entirely empty area of forest, where green and gold mingle. The expanses of wilderness are pierced by railways, roads and motorways. More and more valleys cut into this Cumberland

Plateau. They are accentuated, on such a fine autumn morning, by the trails of fog that linger in them. Now we see a few chimneys rising, their plumes of smoke wafted towards the west. Two rivers converge and on the point that separates them a city can be distinguished; one can discern its business quarter, its factories and, beyond, the dark carpet of its suburbs: this is Pittsburgh.

The plane glides on. The mass of forested area is swiftly replaced by the criss-cross of the grid pattern, the north-south and east-west pattern of roads and the borders of fields: from now on it takes in this area devoted to crop-growing, as far as the Rockies and beyond. The farms are hardly more than dots against the russet and gold background of the yellowed maize and the ploughed fields of autumn. To the right one can make out the Great Lakes.

The picture of the land was transformed as the plane gathered height. At the outset, when the view was oblique, it gave us an unaccustomed vision of the world, but one where objects were still near enough for us to recognize their details. We could see people on the pavements, cars moving, children playing on the lawns. But as our view became vertical, our perception became increasingly distant and disembodied. Colours and shapes helped us to recognize woods, grasslands and crops, but the network of roads and railways was nothing more than a threadlike geometric design. Houses dwindled to dots, and a mighty conurbation like Pittsburgh was merely a dark patch the structure of which was barely discernible. Are we flying over an inhabited world? Yes, of course – the fields, towns and roads are proof, but we no longer grasp directly the seething life and ebb and flow that animates that space. What we discover are the great land masses unseen by the person who remains on the ground: we have seen the Atlantic plain unfold, very narrow at the level of New Jersey, the Appalachian ridges and the Cumberland Plateau, to emerge on to the vastness of the Corn Belt plains, beyond the corridors of the industrialized valleys of the Pittsburgh region.

Today, air travel makes it possible for everyone to gain direct experience of thinking regionally. As we pass from the horizontal or oblique view of the observer on the ground to the vertical view from a height, the things that catch our attention change. From a distance, what is important is no longer this or that house, or pool or wood, but the overall pattern of buildings and the geometry of plots of land, the broad scale. One learns to interpret ensembles that horizontal vision never reveals. But for the mosaic that is visible from on high to tell us the real truth, we must visualize, behind those tiny dots that pattern the huge plains of the Middle West, the farms and the people who live in them, the tall silhouette of the silos where the crops of grain or shucked corn are stored, and the low buildings where pigs and cattle are fattened. Beneath the fog in the Appalachian valleys and the corollas of smoke indicating the presence of factories, we have to imagine mines, their tough immigrant population, the batteries of coke ovens, the steelworks, and the imposing mass of blast furnaces. The tiny spot of the city would not have held my attention if I had not recognized it as Pittsburgh. I know that this site was once a stage on the Indian trails, before being called Fort Duquesne in the time of the French. The little triangle at the confluence of the Monongahela and the Allegheny, whence the Ohio is born, would not have caught my gaze if I had not been able to recognize it as the Golden Triangle the buildings of which have dominated the American iron and steel industry since the time of Carnegie. I can understand this outstanding suburban destiny in the midst of wooded fastnesses now that I perceive the features of the land masses and realize that this great forested island is surrounded by regions teeming with life: the confluence of the rivers and the converging transport routes show me that it would be in the interests of relations across the plateaux to structure themselves around this crossroads.

Regional geography thus rests on a certain way of interpreting the view of the world at two levels; it starts from the ground level, where it notes everything that characterizes the physical and living environment, the infrastructure created by people, their methods for making the best possible use of their lands and subterranean resources, in short, all their activities. It continues by a change of scale which reveals how component parts fit together to form fairly extensive wholes, which are the real objects it describes and explains.

The frameworks in which human activity unfolded were the result of forces that preceded the emergence of humankind; humans had no influence on their level and fluctuations. But as regards most of its features, the stage where today's social actors play their parts is the result of transformations which successive generations have contributed to the environment, and the developments and materials they have gradually accumulated there. In one sense, terrestrial space is a fact that is imposed on mankind; from other points of view, it is the outcome of their enterprise, which reminds the living of the deeds of those who lived before them, and becomes imbued with their hopes. When we study territorial organization, we understand societies through their material foundations, ponder over their ecological basis, give our attention to the infrastructures essential to their daily existence, grasp the ebbs and flows that go through them and give them structure, and pause over the representations and symbols that give places their meaning. It is by way of regional studies that geographers today demonstrate that the discipline they practise is really a science of human beings and society.

Regional studies contribute in some way to answering certain questions that torment humankind. Technical progress upsets the relationships that people have traditionally had with space. The increasingly massive use of fossil and nuclear energy raises output, frees workers from the more mechanical part of their tasks and favours urbanization. New means of transport and communication increase the mobility of goods, people and information. Until two centuries ago, the preoccupation with wresting the wherewithal to exist from a miserly Nature forced men to disperse extensively. Despite industrialization, the principles at work in the organization of space remained relatively stable during the nineteenth century and the beginning of the twentieth. But for a generation changes have been accelerating. The distribution of resources is no longer an essential factor in where people settle. They are increasingly reliant on networks that link them, allow them to specialize and provide them with the goods, information and contacts they need.

Regional geography is indispensable to anyone wishing to understand today's world. For over a century, nation-states have formed the framework in which all forms of territorial organization were inscribed. Their role is called into question by the expansion of flows. Their efficacy is demolished by the increase in comings and goings: to make everything pass through the chief city in order to retain control has become impossible – everywhere bureaucracies are swamped by the overload for which they were initially responsible. The downfall of historical philosophies and national ideologies on which yesterday's political edifices were constructed opens the way to the burgeoning of new identities. Contemporary societies, whether those of advanced countries, ex-socialist countries or those of the third world, are confronted with a gigantic restructuring of space: towards what forms does the organization of space tend today? What are its essential levels? How are responsibilities and powers to be allocated among them? These are the stakes in today's planning policies.

This book opens with a brief sketch of the evolution of ideas in the field under study (chapter 1) and then breaks down the logic behind the regional approach and presents the typologies to which it has led (chapter 2). It then details, in a more technical manner, the cartographical and statistical procedures of regionalization (chapter 3).

The five chapters of Part II review the forces at work in the organization of space, those that divide it into cells and those that cement the component units within larger ensembles. Ecological processes (chapter 4) and economic mechanisms (chapter 5) play an essential role in the segmentation and structuring of space. Social and cultural factors (chapter 6) would appear less decisive were it not for their influence on regional awareness and feelings of identity (chapter 7).

Territory is simultaneously the result, the background and the stake of political action, which calls for specific attention (chapter 8).

Part III presents the history of territorial organizations and analyses their various forms: it begins with archaic societies (chapter 9) and traditional societies (chapter 10), assesses the changes that have happened since the beginnings of the industrial revolution (chapter 11), and, finally, those that are being brought about by contemporary economic globalization (chapter 12).

PART I

THE REGIONAL APPROACH

1

The Development of Regional Studies

1.1 Greek Origins of the Regional Concept

Curiosity about what is happening beyond familiar horizons is universal. It encourages exploration and ensures the popularity of travel accounts. Homer set the seal on this type of work at the very origins of Greek literature: *The Odyssey* recounts the voyages of Ulysses on his way home from the Trojan War. It matters little whether the episodes narrated refer to real places or belong to some mythical space: the journey is still beguiling because it exploits the appeal of the unknown.

The travellers speak to us of what they have seen with their own eyes. We share their wonder, astonishment, or disgust. But their testimony is confined to the journeys they have made and the places where they have stayed. What is happening to either side of them they know only through hearsay – hardly more, truth to tell, than those who have not travelled a step. Their tales leave one dissatisfied; it would be preferable to know a whole land and not just a collection of isolated spots.

The investigations – *historia* in the etymological sense – of Herodotus mark a new departure. He travelled and saw the things he described. But, as Christian Jacob has demonstrated, his description is backed by a new instrument, the map, linked with the development of astronomy and geometry in the Ionian cities. Those representations were still far from perfect; they were not the result of proper surveys, but of a logical reflection that led to the classification of places in ensembles with geometric shapes. When Herodotus described the world, he had before his eyes or in his head, a map of the land as it was known at that time. As a result, his analysis took as its starting point the distinction between the three continents, Europe in the north, and Libya and Asia in the south. The Mediterranean and the rivers Phasis and Araxes separated our continent from the other two. Within each of these land masses, individual geometric shapes could be distinguished, such as the Scythian square, bordered to the south by the Pontus Euxinus (Black Sea), to the east by the Palus Maiotis (Sea of Azov), and by lines parallel with those two shores to the north and west.

Description of the world was no longer confined to presenting the routes of journeys: it distinguished between surface areas which it named and described. In the case of Herodotus, the *a priori* geometry of the basic cartographic document, without which the regional description could not be imagined, limits the value of his analysis. The features that interest us nowadays have more to do with peoples and their ways of life than landscapes and areas: through the evocation of lifestyles, there is no lack of geographical data. Prior to modern archaeological progress, what would we have known of the nomadic habits of the Scythians without the testimony of Herodotus? One forgets the essential methodological contribution of this approach: the passage from the route to the area, to surfaces, their divisions and their boundaries afforded by the projection of earth on a map.

Maps permitting the recognition of differentiated areas made considerable progress in Alexandria during the Hellenistic period, in the atmosphere of research that gravitated around the museum and library. The mathematical representation of the universe was established: the vault of heaven was conceived of as a sphere in the centre of which was the earthly sphere. Both were connected around the axis of the poles and the line of the equator, which was at right angles to it at the centre; they were divided according to the same principles as those revealed by the apparent movement of the sun: the tropics and polar circles formed the boundaries of habitable space.

Each place was defined by its coordinates, latitude and longitude: henceforth, maps would be based on transferring measurements to a base showing meridians and parallels. The principle of modern cartography was attained and it was already possible to establish latitudes. For want of chronometers to establish differences in time between places, longitudes eluded precise determination for a long time, until the eighteenth century in fact. In order to estimate them, there was no choice but to make do with the testimonies of travellers.

Against such a background, the geographer (it was then that the word appeared) was both a geometrician and a scholar. A man who belonged in the library, by making a critical examination of the assessments of his predecessors and the more recent information from travellers, he was able to refine his knowledge of the details of the inhabited world, the *oikumene*. From Eratosthenes (*c.*275–193 BC), who codified the new methods of procedure, to Ptolemy (*c.*AD 100–178), who established an impressive picture of the known positions, progress was considerable, but within the limits imposed by the measurement techniques of the time.

Maps were prepared, the contours of lands and seas, the courses of large rivers and the lines of mountains drawn in: division of the world into regions was then easily carried out. It remained to describe them. In order to do so, the geographer resorted to the testimony of travellers (he was a compiler in this second sense). Strabo (63 BC to between AD 21 and 25), the geographer of antiquity who took the art of regional description farthest, put it very well: 'laying their trust in those sorts of organs of the senses embodied by the various individuals who on random journeys have seen various lands, [the geographers] reconstruct in one single plan the aspect of the inhabited world in its entirety' (Strabo, II.511, C 117). The ultimate aims of this pursuit were many: first, it responded to the impartial curiosity of readers. It supplied an abundance of details about the civilized part of the inhabited world, the best-known part, which coincided with the Roman Empire and, beyond, the Parthian Empire and India, and skipped more rapidly over the fringes of the *oikumene*. It was also intended to assist those who were responsible for the Roman Empire to gain a conceptual grasp of the territory they had to administer: as Claude Nicolet has shown. Augustus tried at the time to form such knowledge by having an overall map of his possessions drawn up and launching a policy of land surveying and enumeration. But the results of those administrative operations, which were very useful for supporting a description with exact data, were still inadequate and would continue to be so until the end of the seventeenth century!

Strabo isolated entire large areas and defined them by their shape; according to him, Spain brought to mind an oxhide with the neck on the Pyrenees, and the Peloponnese recalled a plane-tree leaf. In continental areas he applied divisions along the courses of rivers and lines of mountains. For want of an administrative division, he could often only identify areas by the name of the peoples or tribes that lived there.

There were at the time some precise mentions of topography and climatic features. Towns and ports were named and their most outstanding activities indicated. But the information at the geographers' disposal was slender. The readers they aimed at had no curiosity about economic matters. Developments were devoted to peoples and local history. But pride of place went to the myths attached to places or to the origins of the inhabitants: did they not allow the lands mentioned to be situated once again in the areas of the great stories that lay at the foundations of Greek culture?

Thus geography enumerated the parts of the world. In this way it contributed to the education of the cultured person, which explains the form taken by this discipline in the *Periegesis* of Dionysius of Alexandria, a didatic poem of the second century AD which gives us a rapid overview of the world:

> At the extremity of its [Europe's] point and near the Pillars [of Hercules] live the valiant Iberian people, along the length of the continent, where the Ocean of the North spreads its icy flow, where live the Britons and the white tribes of the Germani, who are ardent worshippers of Arès,* running along the crests of the Hercynian forest. People say this land resembles an oxhide. Farther on, the Pyrenees and the houses of the Celts, near the sources of the Eridanus [Po] with its pretty course: in former times, beside its waves in the solitude of the night, the Heliades wept and wailed for Phaeton; and there, the sons of the Celts, seated beneath the dark poplars, extract tears of gold-flecked amber. Then come the homes in the land of Tyrrhenia: to the east of this appear the beginnings of the Alps: in their midst, the waters of the Rhine rush impetuously to the extreme border of the North Sea.[1]

Geography, conceived of in this way, emerges as a vision of the world where the large areas are clearly situated in relation to one another. Regional description, which at its best, as in the writings of Strabo for instance, posed problems and through the configurations mentioned above made an attempt at explanation, and was reduced to a disparate mixture of features in which the framework of the relief and elements of hydrography alternated with the enumeration of peoples and the evocation of mythological memories. The Greeks invented regional geography thanks to the development of cartography, but they failed to give it a more scientific basis because they lacked precise documentation. The subjects on which their culture was centred led them to persist in a hotchpotch which is without interest for us today.

At the Renaissance, through the rediscovery of Ptolemy and the Greek models, it was this geography that was brought to light again. As in antiquity, it set in motion a cartography that, for the determination of longitudes, relied on an appraisal of the accounts of travellers. The compilation of their impressions and notes on their journeys also served to extend the surface area covered by the meticulous observations for which we are indebted to them, and to fill in the blank spaces on the map.

1.2 The Reformulation of the Objectives of Regional Geography in the Eighteenth Century

The conditions that hampered a description of the earth altered during the eighteenth century. Since the Renaissance, cartography

* The Greek god of war, equivalent to the Roman god Mars.

had made continual progress: the techniques of land surveying and of triangulation led to a strict measurement of distances that separated places and consequently resulted in an exact knowledge of the longitudinal extent of the continental areas. The chronometer, perfected by John Harrison in the 1740s, at last made astronomical determination of longitude possible: henceforth maps were drawn scientifically. There was no longer any need to haunt libraries and consult travellers' tales: it was by means of land surveys that teams, funded by modern states, drew up documents on which everything was transferred with precision to the grid of meridians and parallels. Geographers no longer had to concern themselves with the localization of the phenomena they observed: geodesy and cartography took care of that. To the small-scale documents were added increasingly detailed large-scale maps, such as Cassini's map of France, on which relief is shown, all inhabited places included, as well as wooded zones, vineyards, highways and minor roads.

In areas where a formal administration had been set up, the description of space was made easier by the existence of stable territorial divisions: in France it was a matter of provinces, *généralités* (financial districts prior to 1879), bailiwicks or dioceses. Within these districts the government proceeded to take censuses and counts, as the Roman Empire had done at the height of its power. The birth of statistics provided geographers with new quantitative bases: from then on they could make use of reliable demographic and economic data. The first systematic reviews were carried out in the second half of the seventeenth century. The practice became general in the following century: Spain and Italy experienced some early developments in this field; France and England followed suit at the turn of the century. In the eighteenth century advances were most notable in German-speaking countries.

Regional descriptions were henceforth able to become more exact and richer: dry enumerations of towns or tribal groupings were replaced by the presentation, often accompanied by maps, of districts with clear boundaries for which pertinent data were available. The amount of objective information increased. Nevertheless, the custom endured of referring to outstanding historical features, such as the great men or saints who had made a place famous.

This kind of document was useful to the authorities and of interest to curious readers; it spread and multiplied everywhere. Hardly had the United States declared their independence when Jefferson[2] produced in his *Notes on the State of Virginia* a picture very much in the contemporary taste. But such works did not always meet all the expectations: there was too much of a gap between those born of curiosity about the unknown and those that revealed the administration's concern to gain a good knowledge of the area under its control in order to mobilize the resources and ensure the happiness of the people whose destiny it steered!

The Age of Enlightenment was passionately interested in natural sciences: behind the marvellous complexity of reality, people sought to fathom the designs of the Creator and give prominence to the laws that govern the working of the world. Botany and zoology began the inventory of the vast diversity of living things. Rocks had a history: some bore the marks of the heat, pressure and fire that created them, and others retained in their stratification the proof that they were laid down in estuaries, lagoons or seabeds. The overused words of ordinary language were no longer adequate to refer to the types of rocks, the qualities of soils, and the variety of fauna, flora and climates. To speak only of settlements or houses was to forget the infinite variety of human dwellings. Describing the surface of the earth was to make note of what gave each place its specific character: this might derive from features that were encountered only there, or from an original combination of factors that were present elsewhere, but in different proportions.

The intellectual sensitivity of the end of the eighteenth century encouraged a grasp of reality with more realism, nuances and precision than had been the case until then. Seekers after knowledge had at their disposal a rapidly growing mass of classified and inventoried observations. It was the era when the systematic classification of plants, animals, minerals and rocks was being developed: geography was transformed and became the systematic classification of regions and landscapes. Certain writers laid claim to that transformation: Bernadin de Saint-Pierre deplored the fact that he did not have at his disposal an adequate vocabulary to convey the shapes of reliefs and the shifts of landscapes. Rousseau condemned pedagogies based on discourse that did not recognize the irreplaceable value of experience. The geography being developed at that time did not confine itself to pinpointing places. It sought to grasp what made them differ and to render landscapes distinguishable; drawing was a precious contribution to this process, and engraving and lithography allowed drawn descriptions to be widely distributed. The first task Alexander von Humboldt set himself on returning from his voyage of discovery in Hispanic America was to publish the *Vues des Cordillères et des Monuments des Peuples Indigènes de l'Amérique.*[3] Far better than a lengthy discourse, the magnificent plates in this work show us the natural setting and the marks the Indians had left on it. For Conrad Malte-Brun, the most celebrated geographer in France during that period, geographic description could no longer be content with the dry form and administrative style that had characterized it until then. It had to make the reader share the emotions of the traveller discovering a country for the first time.

To satisfy a public passionately interested in new discoveries, one of the aspects of the geography of the period strengthened a determination to convey the subtleties of the world and to remain faithful to atmospheres. Certain geographers sought above all to underline the picturesqueness of the areas they were describing and the strangeness of their inhabitants' ways. What mattered to them was the physiognomy (the word made its appearance at that time) of the areas described and not their boundaries. Good administration presupposes a consideration of division into formal districts: this was the other aspect of contemporary geography. Divisions inherited from the feudal past allowed the gathering of data on people and resources. But the irregular shapes of the provinces and other land units seemed too arbitrary to suit sound management by the state. That provoked a trend of research in which administrators mingled with naturalists. What they hoped to find was a principle of division that would elude the caprices of those in power: it had to be rational. Three solutions presented themselves: a uniform geometric grid, divisions based on the natural environment, and those founded on the habitual movements and convenience of the people.

The first formula was put forward several times in Europe but was ill-suited to areas where the population was already *in situ* and which would gain nothing by being enclosed within too rigidly geometric a mould. It was the ideal for new lands where the population arrived only after the creation of the administrative framework. Thomas Jefferson proposed it in 1784 for the western territories of the United States. In France, it was the attraction of towns where the new administration would have its headquarters that served as the criterion for the Constituent Assembly when it proceeded with the new divisions into *départements*. The concept of natural regions had no immediate application but aroused the conviction, especially among geologists, that the world was made up of discrete ranges of distinctive areas corresponding with rock outcrops, which were themselves connected with types of relief and soils that influenced the life of the populations. They carried such real weight that, locally, people had known them for a very long time and had given them names. The *pays* frequently corresponded to the units given prominence by the naturalists.*

The way of understanding how the parts of the earth's surface hang together thus made considerable progress at the end of the eighteenth

* The term *pays* has no exact equivalent in English. In French regional geography it refers to a distinctive assemblage based on the physical environment and its human response. It is thus a characteristic cultural landscape with more or less defined boundaries and a strong sense of identification on the part of its inhabitants. The French term has thus been retained throughout the translation, signifying both landscape distinctiveness and the sense of belonging.

century: the maps without which it would have been impossible to represent areas of land were henceforth drawn up by scientific methods; everywhere that modern states became established, available data were enriched by demographic and economic information. Instead of defining the originality of each geographic compartment by ethnographic features, historical memories or reference to local heraldry, they were approached from the starting point of the perceptible realities of relief, rocks, types of life and activities – through the landscapes, in effect. At the same time, people learned to trace boundaries that owed nothing to the ups and downs of political life, but revealed profound natural or social realities. Concern to be descriptive remained paramount: this was the weakness of the works undertaken at that time.

1.3 The Classic Phase of Regional Geography: End of the Nineteenth, Beginning of the Twentieth Century

At the end of the eighteenth century, geographers still did not understand what was to be gained from the idea of a natural region: their attempts at description were contained within rather fuzzy boundaries, while naturalists did not know how to make use of the demographic, social and economic data that characterized the frameworks they had learned to separate. Statistics, which led to this implementation, remained arid and administrative. Alexander von Humboldt, the most accomplished naturalist of the geographers at the beginning of the nineteenth century, sensed the importance of the division into zones of latitude or altitude and made clear the influence of relief on climate and human activity. In spite of all that, he never achieved regional analyses in which all those elements were integrated.

However, geography was making enormous strides. In the space of a century, exploration of the interior of continents was successfully carried out, and observation increased in all latitudes. The number of countries endowed with a regular administration grew constantly: a large part of America joined Europe in this. Colonization provided lands the societies of which were still archaic and imposed the control and organizational structures they had previously lacked. Large-scale topographical maps existed from then on in the developed countries. Towards the end of the century geological maps on the same scale were added. With the aid of these documents it was possible, by traversing several itineraries, to acquire some idea of the relief and landscapes of the whole area: the generalization that allowed passage from direct observation to a regional vision was henceforth within reach of the solitary traveller. To work as a geographer it was no longer necessary to haunt the study or library. In France, Vidal de la Blache and his students decided that in the future the discipline would be practised first and foremost 'in the field' in contact with perceptible realities. That was a decisive breakaway in the history of the regional approach.

People were acquiring a better knowledge of the earth, and could comprehend more directly its surroundings and inhabitants by means of more vivid testimonies. Photography played a great part. The quality of regional descriptions constantly improved. The work of Élisée Reclus bears witness to that: thanks to the abundance of documentation he could utilize, and the network of correspondents that he created worldwide. In under twenty years he compiled a *Géographie universelle*, which remarkably clarified the major features of regional structure in all parts of the world. The work suggested interpretations and stressed reciprocal influences, but its aim was still descriptive. It was the crowning achievement of the tradition born in the eighteenth century.

The perspective changed at the end of the nineteenth century: with Darwin, evolutionism predominated, and it prompted geographers to ask themselves fresh questions. They no longer sought only to give a faithful description of the diversity of landscapes and peoples, but attempted to explain how natural conditions influenced forms of life, and human societies in particular.

From this perspective, the idea of the natural region became central. The task of this discipline was unachievable because it was limitless if each place differed entirely from the others. There were, in fact, whole areas where conditions of relief, soil and climate were sufficiently uniform to be regarded as homogenous. From then on it was a matter of defining those areas, pinpointing them and discovering how groups of humans coped with the constraints of nature there, overcame them, adapted to them, and turned these natural regions into natural human environments.

Geographers at the end of the nineteenth century thus achieved the synthesis of the two traditions of regional analysis that had emerged from the end of the preceding century: the one that attempted to construct the specific character of each place or each complex, and the one that concentrated on demarcating homogenous areas. During the 1830s and 1840s, the way in which basic spatial units combined had given rise to reflections conducted independently by historians and geologists: the resulting ensembles had a personality that ensued from their composition; their unity arose from the complementarities and exchanges brought about by the juxtaposition of contrasting elements. The geographers of the end of the century continued this type of research.

Vidal de la Blache, who gave regional studies their classic form, was inspired by the teachings of Michelet, Elie de Beaumont and Dufresnoy. Analysis of ways of life formed his fundamental contribution; he showed how groups made use of the environment in which they had settled. In the simplest instance, they drew from it everything that was essential to collective living. When that was not possible, or their needs became more varied, trading allowed them to overcome the constraints they encountered in one place: they gave up producing what was not suited to the areas in which they lived and developed activities for which they were well situated. Trade and markets allowed them to make better provision for themselves, within a larger area.

Regional geography thus gave prominence to the connections between areas of land, on several scales: it referred to *pays*, to regions and their combination within national borders, and brought out their 'personality'. Those who engaged in regional geography emphasized the specific nature of areas, referring to the dominant natural conditions and the lifestyles of the men and women who lived there. For those who had understood Vidal de la Blache, there was never any question of regarding each unit as a block closed in on itself: it drew part of its specific character from its inclusion within larger spaces. Through trading activity, people found themselves faced with new ideas, adopted new models and modified their way of seeing things and their manner of taking advantage of the environment.

At the beginning of this century, therefore, regional analysis seemed to be the core of our discipline. It gave geographers a subject for study and caused them to take part in one of the great scientific debates of the period, that of determinism. It also provided them with interesting points of view on some of the problems being debated by society in their time. The regional approach as conceived by Vidal de la Blache allowed an understanding of the profound solidarities that fashioned the nation: the latter was not a modern form of tribalism, but revealed the intelligent will of varied populations to make the most of complementary abilities and talents in order to build a common future.

French political tradition was one of centralization. The political and military advantages that the country had derived from this had long been emphasized, but some people were beginning to wonder if it were not being paid for too dearly in other fields. Regional geography took an active part in the debate that perturbed French society before, during and after the First World War: was it not necessary to review the distribution of powers between the central government and territorial communities? Did the *département* form a framework that was adapted to the needs of an economy based on increased trade, in which mobility had grown greatly? Was not the size of the Parisian conurbation to blame for the country's demographic weakness, because of the low fertility of the women in this great metropolis and the example they set, which was already diffusing to a large proportion of the country?

The success of regional geography appeared to be total in the years following the First World War: the discipline owed its unity to regional geography, for the latter gave it a direction. France, where the art of regional analysis had been carried further than anywhere else, owed the undisputed influence of its school of geography to the work in this field.

The practice of regional geography as conceived by Vidal de la Blache raised some problems, however. The formulae on which it rested were not applicable equally well everywhere. In the still underdeveloped countries of the tropical world, cartography was deficient and statistics were missing or lacking in significance. Practical experience of the terrain had to be combined with perusal of memoirs and accounts of journeys by those who travelled the territory; or one could also draw on administrative reports, where they existed.

In the developed world, the difficulties arose from economic changes that reduced the rural population, increased the proportion of town-dwellers and strengthened the importance of industry and services as generators of employment. The tool offered by an analysis of lifestyles lost its efficacy in a world where everyone adopted the same ways of spending their time, the same kinds of living conditions and consumption; production increasingly eluded researchers who refused to abandon the landscape and visible activities. The study of circulation and the role of towns, though brilliantly carried out by Vidal, was insufficiently advanced to cope with these new developments.

More serious was the change in the intellectual climate. Evolutionism ceased to be at the centre of battles of ideas. Geographers soon realized that it was impossible to establish simple and uncontested relationships between centres and the groups they supported: they spoke of 'possibilism' in order to describe this complexity. The emphasis laid on the region had made it possible to take a good look at human–nature relationships. But though these ceased to occupy the foreground, there was no longer any reason to give preference to the synthetic approach and constantly to mingle the analysis of physical facts with those of human events. Theses in regional geography henceforth concentrated on morphology, rural life or industry: the first neglected the human population, the others treated the environment only superficially. In opposition to regional geography, systematic branches developed: regional studies were often no longer intended to do other than throw light on general problems starting from exhaustive case studies.

The great debates that rocked western societies changed. In France, a demographic renewal took shape, starting in 1942. The fundamental problem was no longer one of promoting fecundity. The idea was spreading that bold policies could speed growth: consequently, no one could bear any longer to see part of the population remain sunk in poverty or mediocrity. Development became the central problem.

The regional issue thus reared its head again in another context. It explains the interest of geographers, beginning in the fifties, in all facets of economic life: gone were the days when they had been basically ruralist. Their interest in the phenomena of circulation and trade became more explicit. Two major factors contributed to the economic differentiation of areas: the diversity of natural endowment and the obstacle caused by distance to the movement of goods, people and information. The impact of distance was minimized by making the greatest possible amount of flow pass through the best-equipped routes and causing all information to converge on the major centres of transfer. Research into urban networks, polarization and regions centred around strong cities permitted the examination of this long-neglected aspect of reality: the article devoted by Étienne Juillard to 'The Idea of the Region',[4] in the *Annales de Géographie* in 1962, was highly symptomatic of this development.

The new geography, as it was formed at the end of the 1950s, stressed this new development: it was based on spatial economic theory, which provided conceptual frameworks to explain, in an open economy, agricultural and industrial specialization, the formation of centres or industrial regions and the architecture of the urban network. The contrast between the centre with a complex economy and the more specialized periphery of national areas, or between industrialized countries and the developing world thus found at least partial explanations.

Despite the progress achieved in analysing the organization of space, regional studies no longer had the wind in their favour: once understood, the logic leading to the organization of circulation flows around hierarchized central places was repeated indefinitely in regional monographs. Around 1970, this type of study seemed heavily under threat: it virtually disappeared from English-speaking countries. In the form of reflection on the spatial organization, regional research lived on in France, but those who practised it became fewer and fewer.

1.4 The Revival of Interest in Regional Studies

The signs of revival have been evident for more than a decade. In France Roger Brunet[5] elaborated a vocabulary of the basic forms of organization, *chorèmes*; this permitted him to free the regional approaches from the descriptive sphere to which they are generally confined. He hoped in this way to extend the works of the French school by systematizing them. He animated within the Reclus Group, a team of young colleagues who launched the editing of a *Géographie Universelle*, published since 1990, constructed on the model popularized by Elisée Reclus at the end of the last century and that was illustrated by the disciples of Vidal de la Blache between the two World Wars: this has relaunched reflection on the forms of regional organization. In English-speaking countries, where the eclipse has been more marked, the success of such themes as place, locale and locality bore witness to new attitudes. Anne Gilbert[6] in 1987 presented an overview of the new developments of the regional theme in French-speaking and English-speaking countries. The frequency of publications has grown even further since the appearance of her short synthesis.

What are the reasons for this renewal? Some are the result of the discipline's internal re-evaluation. The majority of geographers began to draw nearer to the social sciences during the 1950s and 1960s. Breaking with the naturalist spirit which had hitherto inspired them, they adopted a concept that was fashionable at the time among sociologists, often referred to as neo-positivist. The major concern was to demonstrate a scientific approach, which explains the sudden interest shown by geographers in quantitative methods. Initially, the specificity of social facts scarcely interested the majority of them: they accepted the idea that societies are complex constructions, difficult to dismantle but subject to laws similar to those operating in the physical or natural domain. Thus traditional regional geography became devalued since it was purely descriptive. Regional monographs disappeared from the Anglo-Saxon world and became rarer in France. Attention was focused on the forces leading to the organization of space according to universal templates.

By conceiving society as a machine, the accent was placed on the adjustment mechanisms of individual decisions. It explains the regularity observable in many fields: – (1) To ensure an easy switching of communication circuits for example, networks structured around poles offer such advantages that human establishments are almost always organized around these central places. These became hierarchized as a function of the range of services they provide: spatial organization into polarized regions thus offers a general interpretation. (2) The role of scale economies and external economies favours the central areas of industrialized nations, attracting much of the industry and higher-level services. Two types of territorial constructions result: economically diversified regions that dominate and direct national activity and structure the overall national space, and peripheral specialized regions. Thus a general geography of territorial ensembles is elaborated, the essential results of which have been distinguished by Claval.[7]

People are not robots. By considering only the regulating mechanisms, one overlooks the diversity of human beings, their sensitivity and the element of daydream which is always part of their lives: a number of aspects of observed distributions thus escapes explanation. In the Anglo-Saxon world, the radical movement inspired by David Harvey[8] recalls that justice, freedom and equality form a part of human aspirations and that one cannot evaluate in the same way those spatial structures favoured by oppression and those permitted by the full expression of humankind. At the same time, the humanistic approach produced rebirth of curiosity for the sense of place and the personality of *pays*. French geographers insist with Frémont[9] on the life–space dimension of regional existence.

The rise of social and humanistic preoccupations was translated in France into an enriching of the new geography: the notion of threshold limits of flows and of capacity limits for networks, conceived within the framework of space economy, were applied to the flows developed with ecological pyramids, on the one hand, and to social and political relations on the other. The forces that organize space are varied: alongside production, exchange and consumption of goods and services, it is necessary to consider the networks of institutionalized relations which frame social life[10] and the links with power.[11] Governments must oversee the space under their authority: they achieve this by dividing it into districts controlled by a head town: to understand these forms of organization it is sufficient to generalize Bentham's *Panopticon* model, as was done by Foucault.[12]

The idea of applying the systematic approach to the study of regional organization was very popular in France and Italy: first proposed by Dumolard,[13] it is central to the work of Dauphiné,[14] and holds an important position in those of Alberto Vallega[15] and Angelo Turco.[16] Franck Auriac[17] presented a regional monograph of the Languedoc region within this framework. The idea of presenting the region as a system is seductive, but the undertaking requires such detailed data that the theme lost ground quite quickly.

From the 1980s onwards, there was renewed interest in the diversity of the world. The renewal of regional studies took place within a double movement: – (1) The first extended the research conducted over the previous twenty years into systems of organization. It emphasized the impact of the revolution brought to bear on networks and human relationships by computer science and telecommunications: the rapid globalization of the economy led to a generalized restructuring of many spaces.[18] (2) At the same time, a reaction became apparent against over-mechanistic models of man and society fashionable at the time of neo-positivism. The theme of life-world* experienced

* The French term *espace vécu* is difficult to translate exactly. It implies the personal space experienced on a daily basis as a physical and functional setting, but also in terms of a mental perception of its content. The English term 'life-world' has been consistently used to convey this construct.

considerable popularity, as demonstrated by Berque.[19] Questions were posed on the meaning that people confer on nature[20] and the manner in which they develop space: the latter is no longer seen simply as a support for productive activity and an obstacle to movements and exchange. It was seen as having values and an ontological status which varied from place to place:[21] here it is profane, there it is imbued with sacred value. In certain places, nature is feared; a little further away it is considered as the unique source of well-being. The study of spatial differentiation became enriched by taking into account the dimensions attached to places by human imagination, and by dwelling on the way in which man lives in his familiar setting, aspires to other horizons or isolates himself from contacts or other relations. Drawing on the transformation undergone by southern Scania at the end of the eighteenth and beginning of the nineteenth centuries, Allan Pred demonstrates in *Place, Practice and Structure*[22] how to marry the facts of social structure which explain the organization of space and its changes, and awareness of the existence of the dimension of regional facts.

The interest in regional studies was stimulated by the disappointment with radical approaches, almost always based on a Marxist point of view: they claimed through Marx to understand universal laws governing human societies; this gave them their confidence. But then it was discovered that it was not enough to state that a society's production system was capitalist in order to explain the location of its activities and to foresee its dynamism. Even for those who continued to believe that the profound truth of our societies lay in the connections described in *Das Kapital*, a link was missing. General forces combine in unpredictable ways at the levels of towns or regions: the accumulation of capital creates, where it takes place, rigidities that incite dynamic entrepreneurs to look for other locations. In *The Limits to Capital*[23] David Harvey thus introduced into Marxist reasoning the spatial dimension which was lacking. The study of places becomes essential to understand local synergies. The renewal of regional studies willingly assumes this aspect in Anglo-Saxon countries: a work like that which Derek Gregory devoted to the industrialization of Yorkshire[24] demonstrates this.

Subsequently, Derek Gregory has shown in 'Areal Differentiation and Post-Modern Human Geography'[25] that the mutation affecting the economy inspired by Marxism is parallel to those concerning social and anthropological studies. In sociology, Anthony Giddens[26] attempted to construct a theory that took account of extent as one of its constituent dimensions: traditional approaches were thus relativized because they could not take into account the totality of reality. This depends on forces that are not uniformly distributed. It was to stress this that Giddens introduced a new term,[27] that of the 'locale': the definition that he gave to it unfortunately was too vague for it to be employed in the field. The impossibility of translating the term into French also underlined this difficulty. In anthropology, the techniques of exhaustive description, of 'thick description' of Clifford Geertz, also enabled the capture of what escaped from generalizations and expressed the fullness of places.

Derek Gregory considered then that the re-found attention to places and regional differentiation of the earth translated a profound evolution in the social sciences – which many associate with the idea of postmodernity. It was between 1986 and 1989 that the impact of the essay by Fredric Jameson on 'Postmodernism, or the Cultural Logic of Late Capitalism'[28] was greatest: one sees it in the publications of Harvey,[29] Soja,[30] as well as in Gregory's paper. The last, however, is the only one clearly to draw from the theme of postmodernity the idea that geography is indispensable to new forms of knowledge, because it permits us to explore the local dimension of the combination of forces at work in society.

Nicholas Entrikin took a further step forward in the debate in *The Betweenness of Space: Towards a Geography of Modernity*.[31] Here, he shows the evolution of the position accorded to place and to the idea of the region in the epistemologies that have succeeded each other in geography since the eighteenth century. Geography for a long time applied in this domain a decentred viewpoint, that of the external and objective observer. It is fashionable to criticize such an attitude – but Entrikin recalls what this has contributed 'in spite of the present success of scepticism as far as decentred rationality is concerned, the value of a decentred theoretical perspective cannot be doubted as much in the human sciences as in the natural sciences' (p. 132). More recently an effort has been made to renew the regional approach in geography. It is characterized by 'a greater willingness to go beyond traditional "facts" relative to places to examine what is more subjective in the experience of places' (Entrikin, p. 133). Entrikin thus establishes that there are two possible dimensions in the regional approach. The apprehension of the second is linked to the judicious use of narration, which obliges geographers to reflect on the rhetorical procedures to which they have recourse.

The return to the region is very active in Anglo-Saxon countries. Sometimes it is linked to the concern not to neglect an essential level in the mechanisms of the contemporary world: this is what encourages Phil Cooke in *Localities: The Changing Face of Urban Britain*[32] to adhere, in the British context, to the local dimensions of present economic and demographic evolution: is it not at this level that initiatives are generated that permit development whereas elsewhere de-industrialization has not found a compensating redevelopment? Since we must understand the role of actions taken from below, localities constitute the most satisfactory framework.

In his book *A Question of Place*,[33] Ron Johnston reflects on the orientations that have imposed themselves on the Anglo-Saxon world since the beginning of the 1980s. He recognizes the importance that the regional dimension assumes in contemporary human geography: it appears indispensable for him to integrate knowledge that otherwise tends to be scattered in all directions. He refrains from returning to the concepts of traditional regional geography and, like Cooke, accentuates the idea of locality.

In this book we propose to present what 'decentred' and 'objective' analysis of the facts of regional organization brings to the understanding of geographical facts and of the problems encountered by past and contemporary societies. We also insist on the existential dimension of spatial connections. Our approach is based on the French tradition of regional study and seeks to integrate within this the most recent developments.

The success of new orientations arises from the problems confronted by our societies. States, since the 1930s and until the end of the 1960s have held an essential place in the economic development of industrialized societies. In socialist countries, states operated command economies: in the liberal economies intervention was by fiscal policies or by credit controls: the concern for social justice led to an increasingly wide redistribution of income: along with health and education services, the State was responsible for transport infrastructure, housing and research and development in certain sectors.

The weight of systems thus created, called into question their effectiveness at the beginning of the 1960s. At the same time, the progress of economic globalization deprived governments of the means of pressure that they disposed of in relation to firms. The rate of growth slowed down in the most developed countries. The policies conceived and enacted at the national scale were demonstrated as powerless to modify the evolutionary trends. Was it not time to reconsider political regulations? Was it not preferable to have a better distribution of powers to avoid congestion in the decision-making centres? To stimulate new initiatives was it not necessary to give local authorities more weapons to defend themselves? The political world wishes avidly to understand local and regional realities. Citizens dream of a more active role in the management of their fate. They no longer believe in the golden tomorrows proclaimed by yesterday's progressive parties that led to the familiar totalitarianisms. People nowadays expect to deal with their lives here and now: what happens at a local scale assumes a new resonance. And, as movements, trade and communications take place on a planetary scale, it seems more and more necessary to protect the immediate world against the trend towards a general levelling.

Regional geography today does not fulfil the same scientific objectives and does not attempt to resolve the same questions of society as yesteryear, but without the concepts and methods gradually perfected since the first descriptions advanced by the Greeks, it would be impossible to satisfy today's curiosity and confront new causes for concern.

Further Reading

The available recent literature on the history of geographical thought is voluminous. The following texts collectively cover the scope of chapter 1 and contain extensive bibliographies.

Agnew, J., Livingstone, D. N. and Rogers, A. (eds), *Human Geography: An Essential Anthology*, Oxford: Blackwell, 1996, 704 pp.

Chorley, R. J. and Haggett, P. (eds), *Models in Geography*, London: Methuen, 1967, 816 pp.

—— and Haggett, P. (eds), *Socio-Economic Models in Geography*, London: Methuen, 1968.

Cloke, P., Philo, C. and Sadler, D., *Approaching Human Geography: An Introduction to Contemporary Theoretical Debates*, London: Paul Chapman, 1991, 240 pp.

Gilbert, A., The new regional geography in English- and French-speaking countries, *Progress in Human Geography*, vol. 12, 1988, pp. 208–28.

Gregory, D., *Ideology, Science and Human Geography*, London: Hutchinson, 1978, 198 pp.

——, Martin, R. L. and Smith, G. E. (eds), *Human Geography: Science, Space and Social Sciences*, London: Macmillan, 1994.

Haggett, P., *The Geographer's Art*, Oxford: Blackwell, 1990, 294 pp.

——, *Locational Analysis in Human Geography*, London: Arnold, 1965, 339 pp.

——, Cliff, D. and Frey, A., *Locational Analysis in Human Geography*, 2 vols, London: Arnold, 1977, 605 pp.

Hartshorne, R., *The Nature of Geography*, Lancaster: Association of American Geographers, 1939, 469 pp.

——, *Perspective on the Nature of Geography*, London: John Murray, 1959, 200 pp.

Harvey, D., *Explanation in Geography*, London: Arnold, 1969, 521 pp.

James, P. E. and Martin, J. M., *All Possible Worlds: A History of Geographical Ideas*, 2nd edn, New York: Wiley, 1981, 508 pp.

Johnston, R. J., *Geography and Geographers: Anglo-American Human Geography since 1945*, 5th edn, London: Arnold, 1997, 475 pp.

—— (ed.), *The Challenge for Geography. A Changing World: A Changing Discipline*, Oxford: Blackwell, 1993, 250 pp.

—— (ed.), *The Future of Geography*, London: Methuen, 1985, 342 pp.

——, Gregory, D. and Smith, D. M. (eds), *The Dictionary of Human Geography*, 2nd edn, Oxford: Blackwell, 1994, 744 pp.

Ley, D. and Samuels, M., (eds), *Humanistic Geography: Prospects and Problems*, London: Croom Helm, 1978, 337 pp.
Livingstone, D. N., *The Geographical Tradition: Episodes in the History of a Contested Enterprise*, Oxford: Blackwell, 1992, 448 pp.
Stoddart, D. R., *On Geography and its History*, Oxford: Blackwell, 1986, 335 pp.
Unwin, T., *The Place of Geography*, London: Longman, 1992, 273 pp.

2

The Regional Approach

2.1 The General Characteristics of the Regional Approach

The unity of regional geography stems from the methods of approach that it uses. It is a matter of making the transition from what the eye can observe in one spot, or along a transect, to a totally different spatial reality, on a scale that is considerably reduced, allowing an overall view of extensive areas. This alternative view is a reality because it is based on knowledge of the terrain, and abstract because it is detached from the complexity of the perceptible world; it is reduced to a few major features, chosen as such – and localized on a map. It is thus possible to avoid drowning in detail and features are captured that elude the ordinary traveller.

The regional approach transcends the direct and simple perception of the earth's surface and replaces it with a constructed view. It combines three elements:

(1) A basic cartographic representation on which observations are recorded; without this it would be impossible to carry out the alteration in scale that transposes points and lines to surfaces, and one would not manage to discover overall distributions.

(2) Direct observation of at least part of the zone studied: without that nothing would guarantee the validity of the representation that is portrayed. The regional approach does not seek only to construct a view on a different scale. Whoever practises it must back up the generalizations he or she advances by personal experience of the area concerned: if not, how could distortions that are always possible be avoided, and how could one be certain that pertinent features have not been overlooked?

(3) The indirect observation made available through the use of data collected by others. This involves the risk of error: data gathered by different observers are not necessarily homogeneous; different biases and distortions may result. Procedures of sifting and checking must be employed to avoid them; but it is absolutely impossible for one person alone to carry out a direct and exhaustive survey of the terrain except in the case of strictly local monographs.

The regional approach always has the same aim: to replace the limited and discontinuous conception of the earth's surface with an overall view, that of the observer who sees things from above. Today, when air travel is within the reach of most people, and aerial photographs and synthesized images provided by remote sensing are available, that may seem trite. We must try to comprehend how much the formation of a cartographic basis and its use as a back-up for a second-hand observation required imagination and effort in earlier times. Modern technology offers facilities to a regional approach that could scarcely have been dreamed of barely a generation ago. It ensures the building and constant updating of the cartographic base which it combines with much survey data. Nevertheless, experience of the actual terrain remains essential. Today's sophisticated instruments of observation yield an image of whole sections of the earth's surface starting with data so rich that the regional approach appears to be undergoing total automation: but in order to interpret the satellite image successfully, one must compare it with the terrain itself; furthermore, the communication flows, invisible yet of such importance in the life of the regions, will always elude the cameras.

2.2 Contexts and the Regional Approach

It is not so easy to implement the regional concept in underdeveloped countries, which lack the established administrative structures, and this was even truer of traditional societies in the past. Formal documents on a large or medium scale are lacking or are difficult to access owing to the secrecy imposed by the state security services. Researchers themselves must carry out surveys; returning to the origins of the art of representing the earth, they must collect the references provided by the itineraries and travel narratives to determine the relative positions of places and rough-out a map of them. Doing this is perhaps easier where vegetation is not too dense and the relief is pronounced: from a hilltop the gaze can take in a large expanse from which it is possible to make a sketch plan. By combining successive viewpoints and with the aid of rapid triangulation, quick and relatively exact schemas can be developed.

Direct investigations raise many problems. Extreme or dangerous physical environments and hostile populations or states sometimes make access to a given zone out of the question. Dialogue with the natives implies coping with difficult languages. Their fragmentation into innumerable dialects further complicates the investigator's task. It is often necessary to resign oneself to working through interpreters – with all that this implies in the way of approximation, involuntary bias and, sometimes, deliberate deceit. Interpretation of the data is all the more difficult because the outlook of various peoples is totally alien. The investigator must constantly proceed by efforts at decentring and recentring. Ideas which appear clear to us, such as those of ownership or location for instance, may have no equivalent in the legal frameworks of the country under study.

The absence or shortcomings of a formal administration deprive the geographer of that basic instrument formed for him by the separation of areas into basic districts, parishes or communes. Nothing tells him on what fundamental network of links society is built. In the end, there is nothing to allow the assumption that it is the same everywhere: the size of communes varies considerably in a country that has been so completely and so long unified as France. So what complications are to be expected in less standardized countries?

The acquisition, status and use of land often appear vague and fluctuating to the western observer. In such contexts it is useless to draw up a cadastre (even if modern cartographic techniques are available) and to pinpoint the state of occupation of the land. Of what value is a document if the lots change frequently and if there are time-limits to private use? How do we make a distinction, where private ownership exists, between land and buildings registered under the identity of those who actually acquired them, and those that have been acquired by a 'front-man'? Local dignitaries are expert at parrying all the whims of fiscal inquiry in this way.

Without an administrative system, there are no regular compilations of data: reliable statistics are wanting. Valid censuses of the population and civil status are not available. No evaluation of production is formally carried out. If investigators are recruited with a view to carrying out counts or administering questionnaires, the coordination of their findings falls foul of the absence of basic districts with known boundaries that are universally accepted.

In developing countries

In developing countries conditions have altered considerably during the last generation. Cartographic coverage is increasingly assured, and regimes which were so mistrustful of foreigners that consultation with them was banned are becoming rare – do not spy-satellites continually keep the Great Powers informed of what is happening in the very heart of national territory? The state exists, and its first concern has been to extend formal administration to the entire country. This has involved the demarcation and often the mapping of basic territorial districts.* The legal population is organized, periodic censuses and countings are effected. Accounts are kept of the principal products, particularly those destined for export.

Difficulties are connected with the margin of error contained in published data. The precision of figures is often illusory and it is impossible to trust them. An individual does not find it easy to exist on the registers of an impoverished country. What is the good of making the effort to go and declare the birth of a baby whose chances of survival are slim (especially if it is a girl)? Assessment of production is even more questionable: administrations reply imperturbably to questionnaires addressed to them, without making the necessary inquiries. Information is recopied from one year to another; a few minor alterations are introduced just to make it appear truthful. Generalized corruption enables the powerful to conceal their wealth. When the researchers, exasperated by the poor quality of the sources open to them, decide to organize their investigations, they encounter endless harassment. The times are over when researchers were allowed to put their questions freely as they wished. Now authorizations must be sought from a slow and pernickety administration. Once they are obtained, it becomes obvious that refusal to reply or systematic lies are frequent occurrences wherever an authoritarian regime is in power. Agents who may be recruited to help are not always trustworthy. Precious time may often be wasted in checking their work.

Nevertheless, researchers could observe things directly, and obtain a few worthwhile pieces of information during the interviews they conducted or through questionnaires administered in person. All would be well if their movements were unrestricted – which is far from being the general case. Mistrust of foreigners or fear of the administration often make it difficult to establish contact.

There is a great temptation to be content with analysing facts which are today easily observable: the importance given to data acquired by remote sensing may be partly explained by this. This kind of survey is not dependent on the good mood of the local government and is in no way affected by the greater or lesser degree of corruption of the agents

* The French text makes frequent reference to 'circonscriptions territoriales elementaires'. This refers to the lowest level of local administration (in France, the commune) which may or may not have electoral significance. The phrase has been translated throughout the text as 'basic territorial districts'.

of public officialdom. Information is supplied on a cartographic basis that permits its direct use. It cannot be used, however, without a minimum of measurements and checks carried out on the ground. What it contributes has to do with relief, vegetation at different times of the year, soils, the effects of erosion, and the forms of occupation of the land and habitat; this is indeed considerable, but leaves aside people, their behaviour, activities, mobility and representation.

In industrialized countries

The conditions in which the regional approach is developed in industrialized countries are very different. Their territory has been the subject of formal and precise topographic surveys so that geographers have available cartographic documents of good quality and on all scales. Moreover, they can consult cadastral documents that combine large-scale representation, structure of the property and, to a certain extent, land use. In addition they obviously have the resources of aerial photography and other forms of remote sensing.

Study of the physical area is facilitated by the accumulation of observations. Many are already in map form. This is usually the case with geological and mineralogical data and, to a lesser degree, with those bearing on soils and vegetation. The range of long-term meteorological observations makes it easy to define average climatic conditions, reveal the succession of its characteristic types of weather and show the fluctuations that affect it. The network of recording stations is sufficiently dense to pick up local nuances and to give some idea of the temperature near to ground level, at a height of one metre, at treetop level, etc.

The administration long ago defined the basic districts for which it collects data. In terms of general censuses one can know the population, categories of age and employment, housing and its varying amenities, as well as daily journeys. More specialized statistics, often gathered at less frequent intervals, reveal the characteristics of farming, industrial and commercial businesses. Other departments record the variations in the flow of goods and information, or of tourist numbers.

In so far as modern societies are close to us and thus seem directly visible, the abundance of data induces a temptation to give up fieldwork: a high proportion of regional studies result from treating and elaborating sets of statistics gathered by the administration. As they are furnished by territorial districts, it is easy to convert them into map form. It is from the regionalization thus effected (we shall see the methods in the next chapter) that the approach can be organized.

The power of analysis available to the geographer working in industrialized countries is increased by the quality of the demographic databases and the wealth of indirect periodical statistical surveys that he can expect. It is easy for him to deal with many data series relating to extended areas. Whereas the composition of the slimmest monograph requires long periods of time in traditional or underdeveloped environments, it is easy in industrialized countries, thanks to the handling of standardized data available for a whole or several nations, to conduct an analysis at more than one scale, to reveal the overlapping of forces involved and launch oneself single-handedly into an analysis of ensembles of considerable size.

There is a danger of letting oneself be led astray by available sources. They are never perfect, and some of the shortcomings which render them unusable in traditional societies are ever present: in the space of two months, a small team of investigators demonstrated that over 100,000 people had been omitted in Paris at the time of the 1982 census! Official departments note only what is of interest to them, which explains the absence of data in areas that are sometimes important. Moreover, information provided must remain anonymous, which is possible only when it concerns fairly large numbers. Certain data are not published at district level. The other danger is of overlooking some of the most significant aspects of reality: we make less effort to be objective in societies that are more familiar to us, but by giving up fieldwork we risk not seeing the forces that are really at work. In this respect there is often a tendency to call on simple mechanisms of an economic nature which fashionable interpretations have popularized: we are unaware of everything that would lead us to real relationships, which are often very different from those one imagines purely because it is less trouble.

So the risks of error do exist, as does the danger of not getting to grips with fundamentals when everything appears to be falling into place in the framework already obtained. Industrialized countries are still a kind of paradise for the regional approach. There is, or was, one exception however – a matter of size: socialist regimes have always reserved access to information for the State and the Party. The Soviets were so suspicious that they introduced systematic errors into their maps in order to make them inoperative, and the scales of those to which they allowed access were too small to allow the inclusion of the results of detailed study. Their sets of statistics were doubly false: because of erroneous declarations made by firms and private individuals who had no interest in supplying information that might rebound on them, and also because of a concern to mislead enemy intelligence services. Direct practice of unsupervised inquiries came under the heading of subversive activities, and foreigners who took the risk were treated as spies.

Developing socialist countries concealed lacunae in their basic infrastructures. It was always impossible to carry out regional inquiries in countries such as Cuba or Vietnam. North Korea refuses, even today, to allow any overtures. To a lesser degree the same shortcomings were to be found in countries that remained on the fringe of the socialist world, Madagascar, Congo-Brazzaville, Guinea and Algeria.

Crisis situations

The difficulties inherent in the regional approach are not the prerogative of under-administered countries. When crises occur, the systems responsible for the gathering of data often tend to dissolve. For that reason, regional geography has long been confined to periods of calm and equilibrium. There was unwillingness to concentrate on a country when a cyclical slowing-down of the economy was paralysing its activities and leaving a large proportion of its adults jobless. There was no question of taking an interest in the periods of difficulty and the upheaval resulting from the fury of the elements or the unleashing of passions at times of great social or international conflict.

But why should situations of balance or slow development be the only ones to deserve geographic study? Scientific knowledge of extreme situations and their spatial configuration is very useful in perfecting effective policies of prevention. The only problem is that scientific work is then revealed as thankless. A volcanic eruption, an earth tremor, the path of a cyclone or catastrophic floods alter the environment and, at least temporarily, destroy or render inoperative for some time the infrastructures of transport and communication, and bring down or damage some constructions. There are many victims, and health services have to take care of the wounded, look after all those who have been traumatized and make sure that the disorganization of networks supplying drinkable water and disposing of liquid waste does not provoke the spread of epidemics.

To study the distributions arising from these upsets, and to capture them at the very moment when the situation is most dramatic, developed countries have available a permanent database, including maps, long-standing physical or meteorological records and often plentiful statistics. But the first victims of the situation are frequently the administrative departments: officials may figure among the dead or wounded; some, seized by panic, may have fled; offices are totally or partly destroyed; telephone connections no longer work, the electricity supply has failed, computers remain mute; in the agitation that overcomes the population, responsibility for authority vanishes. No one thinks about informing the services concerned: their reputation has been tarnished by what has just occurred.

When geographers interested in extreme situations want to set the regional approach in operation, they find themselves in the same position as someone trying to understand an unknown land. Certain fundamental documents are lacking – the maps of the zones affected by the catastrophe and the degree of intensity with which they have been hit. Administrative districts stop playing their role: they no longer gather essential data, or do so in a disarray that prevents one from knowing whether they are complete and trustworthy. To measure the impact of the disaster, it is necessary to work with fragmentary information supplied by fugitives, refugees or journalists and members of the first rescue teams to have visited the sites. If one wishes to obtain more precise data, one must conduct interviews, or have them conducted – which is often impossible in a time of chaos.

If geographers cannot capture the drama when it actually happens, they must reconstruct by means of the testimonies of people they meet and question after the event, and by initiating inquiries that the administration follows up when it is back in place. This is how the effects of the typhoons in Taiwan and mainland China were analysed. Interpretation of the sources is always a delicate matter: witnesses do not always remember everything and have a tendency to dramatize the situation and exaggerate the part they played; statements about disasters are inflated in the hope of obtaining greater compensation.

The very fluid situations that accompany armed conflict and guerrilla operations are beginning to hold geographers' attentions. In particular, they are aware of the displacement and regrouping of the population which are associated with insecurity or revealed by the increase in urban population – for example, in Saigon or Phnom Penh between 1960 and 1975. They find themselves faced with the difficulties of conducting a regional approach in contexts where a permanent organization can no longer be relied on for the collection and recording of data.

The regional approach has a single aim, but it does not always combine in equal proportions basic cartography, survey of the terrain and the use of information that has been systematically collected by administrations. In cases where the lack is greatest, the change of scale that reveals overall distributions is a product of essentially mental gymnastics. The geographer has a picture of the region in his mind's eye and ends by associating specific features with each of its component parts. When one has at one's disposal all the means that the organization of bureaucracies and technical progress have successfully contributed, the operation has less to do than in the past with intuition, flair, and the ever hazardous and difficult exercise of mental generalization. It relies more on cartographic and statistical processes, the systematization of which continues to make progress.

2.3 The Regional Concept and Territorial Framework

The regional concept begins from observations that are very precise or concern limited expanses. From these it draws conclusions that are valid for considerable areas. The quality of the generalization depends on the areal framework in which the elements gathered initially are inserted.

The space where the basic surveys are conducted is never perfectly uniform. It is composed of elements that are repeated so that it may be compared to a fairly tightly-woven fabric or a rather fine-grained leather. The surface being studied is not like the flat calm of a lake in fine weather. It is criss-crossed by ripples or waves: conditions of observation vary according to their amplitude.

It may be that several of these features are always taken in by a single observation: enough pieces of information are then available for one to know the elements that combine and, at a smaller scale, outline overall configurations (see figure 2.1). By transferring these data to a base map, and representing by the same symbol all those that are similar, we see ensembles appear. Between them there are discontinuities or transitional zones. Even if none of the places for which one has observations is to be found on the boundary, we know that the limit passes between the closest locations with different signs and characteristics. When the field data is also obtained from transects, the transition is located with precision: it appears as an abrupt discontinuity or as a more or less long segment where characteristics are mixed in variable proportions.

The vegetation cover in regions where nature has remained untamed is often ordered on this model. Dense forest dominates at certain points, grassland elsewhere. Passage from one to the other can be by direct contact, or by a transitional belt where the forest opens out into clearings, contains trees that are less fully developed and more light-loving species, before giving way to parkland where stretches of grass are dotted with copses and isolated trees.

Land forms have other characteristics (figure 2.2): the gaze generally embraces only the two sides of a valley; when one looks from a higher point, the vista that opens over the rolling summits of the hills creates the illusion of an almost boundless view, but the heights hide the essential features of the area. One can see for a long way, but what is perceived corresponds to only the tiniest part – sometimes less than 1 per cent – of the whole area, and covers only its most salient elements. To arrive at a synthetic view is more difficult than in the preceding case. Generalization remains fairly easy if it is supported by existing maps or aerial photographs. Thanks to these documents, it is possible to pick

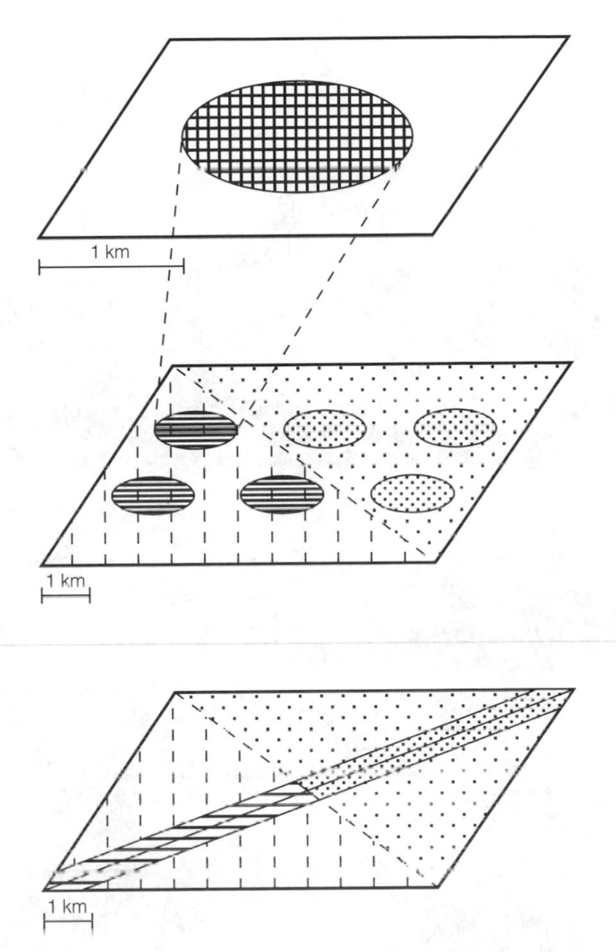

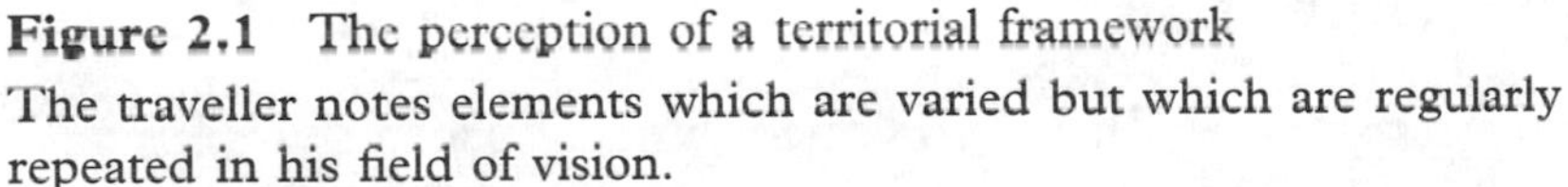

Figure 2.1 The perception of a territorial framework

The traveller notes elements which are varied but which are regularly repeated in his field of vision.

Localized observations are transferred to a map. After reduction in scale, each perceived unit appears homogeneous. The observations are distributed on the map in two groups, one to the south-west, the other to the north-east. One may thus suppose that, at the scale of the map, there are two homogeneous ensembles separated by a frontier.

The location of the boundary is made possible by a transect.

out a few basic forms from within the area under study – dissymmetric talus slopes in the west of the zone, for instance, and parallel ranges in the east. Must the features of the latter be rendered more exactly? To get a fair idea of them, it is enough to have observations available on their summits and slopes, and the valleys that separate them. When topographic surveys, aerial photographs or synthesized imagery are not at one's disposal, the first phase of a regional approach takes longer: it

Figure 2.2 Trace of the observed parts

In A, the relief as it appears to the observer. As he sees the distant crests, he has the impression of capturing a large area.

In B, what he sees has been transferred to a map: it embraces a very small extent and what is unobserved increases very quickly with distance.

implies a direct survey on the ground to demarcate the extent of the basic elements and to classify them. Point-based analysis is replaced by exploration of surfaces: a certain number of transects are established which, for each landform, specify the overall dimensions and, at each site, the slopes and orientations as well as the nature of the

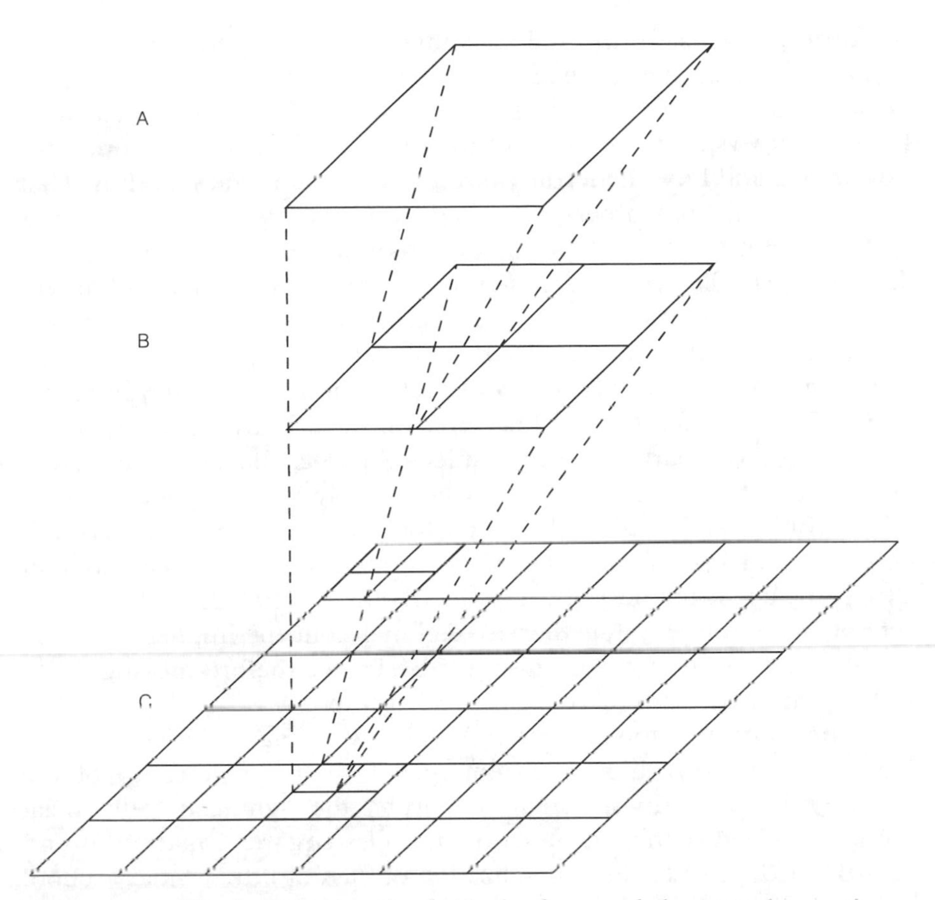

Figure 2.3 The basic units: the field, the farm and the township in the American Middle West

The rural landscape was perfectly standardized at its creation, which gives the observer three privileged scales of observation: (A) the field a quarter of a mile square (16 ha), (B) the farm a half a mile square (64 ha), and (C) the township of six miles square. The field covers a quarter of the farm, and the farm a quarter of the 36 sections of a mile square which forms the basic administrative unit. Two centuries have elapsed since the creation of the grid pattern, the fields have been enlarged, the farms have been reorganized, but the overall regularity continues to facilitate observation.

soil and subsoil. Once the types have been determined, they may be recognized on the basis of a few careful observations, which reduce the number of survey points.

The problems posed by social or economic realities are of the same kind. There are monocultural regions where the land is worked by farms that are uniformly dispersed; such is the case with very broad portions of the American Middle West (figure 2.3). Reality comprises two levels of basic structures: crops and farms; the clarity of both is fine enough to be easily distinguished by the observer. By its general uniformity, the use of the soil resembles what is apparent in the case of natural vegetation: a generalization may be made without risk of error when based on careful observations. The farms also need to be known: surveys must be made of their scale. Without them, one cannot understand how the fertility of the soil is maintained, and by what means (in terms of workers, machines, fertilizers, pesticides, buildings and other infrastructure) productivity is assured. In the case of the American Middle West, the base unit has grown, but remains limited: from 64 hectares originally, at the time when the Homestead Act led to dividing up the prairie and the allocation of plots to colonists, farmsteads passed from 200 or 300 hectares in the more fertile zones, and to 2,000, 3,000, or 4,000 hectares on the arid fringes in the west.

For the most part, the basic units that must be analysed before changing scale are far larger. Even in a purely rural region, methods of cultivation and feeding livestock form only a part of everyone's concerns and use of time. People live in society: they need to visit shops, the bank, the post office or the doctor, to send their children to school, relax, or play together. A regional concept implies that one should know what is going on at this basic level before moving on to higher ones and defining more general pictures.

In the context of the American Middle West (again, refer to figure 2.3), it is the township which emerged from the setting up of the initial grid of land division that constitutes the significant cell: social life is organized within this area of six miles square. Children attend (or attended, prior to the generalization of bussing) the primary school built on a central plot. Churches and chapels, the general store and the garage form the nucleus of this primary centre of social contact. In the countries of western Europe, rural units, the heirs of which are today's districts (in France, 'communes'), were organized around the parish church: the administration understood what was to be gained from retaining this framework as a basic district if it wanted to have a hold over all levels of society. It makes the geographer's task far easier: the organized gathering of data can be done in the smallest area considered necessary before changing the scale.

The situation is altering today. In the last half-century mobility has increased to such an extent that people's daily lives and the elementary forms of social contact are no longer contained in communal boundaries. The proportion of those whose job lies outside their place of residence continues to grow. Shopping is done partly or entirely in commercial centres a few kilometres away, often farther. Associations to which people belong recruit from a radius often exceeding a dozen kilometres. The administration and the Church find themselves encumbered with spatial frameworks no longer suitable to the scale on which today's collective life takes place. France has not yet really tried to modify its basic structures. The majority of the other European countries have done so, but their chosen solutions may not always be perfectly satisfactory.

Before proceeding to data, the new regional geography must therefore determine within which frameworks today's social life is inscribed. British authors lay emphasis on *locality*, which they match with labour catchments. The choice of minimum areas to be borne in mind during investigations thus influences the rest of the regional approach.

2.4 The Results of the Regional Approach

The nature and number of features retained; homogeneous regions and polarized regions

The regional approach teaches one to see what the eyes never perceive directly (or never used to before the coming of aircraft and satellites); it reveals groupings, ensembles, combinations the analysis of which constitutes the very objective of geography. Their nature and structure are varied. Geographers have long since learned how to instil some kind of order by establishing typologies.

The divisions that emerge from the regional approach differ according to the number of features retained in the observation and survey phase. It may happen that only one is retained. Geographers then hesitate to qualify as regions the land ensembles that emerge. They prefer to reserve this term for the divisions derived from an ensemble of features. About fifty years ago André Cholley[1] put forward a more precise vocabulary. He talked of *domains* when the factors dealt with concerned only relief or climate, *milieux* when they related to living forms, and *regions* when they also included people and their activities. This usage did not become general, but the idea is sound. It avoids any equivocation.

Each place provides various characteristics that define it and that can be used to compare or contrast it with others. When one refines the observation carried out with this aim, it is the links developed vertically between the aspects of reality that are emphasized: one notes those links that are established between the climate, the soil, the vegetation, the occupation of the area and the people's ways of living and working. The divisions obtained then after a change of scale delimit homogeneous ensembles.

One may equally record, as pertinent features, the links maintained with the federating nuclei or poles of social, political or economic life – usually the towns. A second type of land unit results: this is termed a polarized region.

The two processes are equally interesting: the first captures the ensemble of the features composing the physiognomy of a whole and may be read in the landscapes; the second lays stress on the comings and goings, relations, commercial life; it gives us information on the functioning of territorial groupings.

Types of region

At a very early stage naturalists and geographers were aware of the existence of divisions linked with the influence of relief, soils and climate, which are easily decipherable from the way vegetal formations are divided. It was to describe this sort of unit that the term *natural region* was coined at the beginning of the eighteenth century.

In terms of human geography, research centred on productive activities showed that they are often grouped in territorial clusters in which they form the dominant feature. Since the beginning of the century,[2] the terms *homogeneous economic region, agricultural region* or *industrial region* have been employed. The logic of services is different; as they must remain within the reach of the clientele, they tend to be divided among central places. Certain tertiary activities, however, form an exception and are not dispersed: coastal areas, mountains and places with old civilizations attract those who wish to live an open-air life, enjoy themselves and change their horizons. In such places, therefore, one may see a build-up of hotels, camping sites, casinos, golf courses, and the like; these may be called *tourist regions*.

Not all *economic regions* have the same composition. Some are extremely *specialized*. Others have varied activities. It is thus necessary, alongside the dominant activity, to take note of the presence or absence of other orientations. If this is not done, very different ensembles may be classed in the same category. In the United States, there are regions that subsist on tourism alone. Once the season is over, they empty almost completely; only maintenance and security staff remain. Conversely, the Côte d'Azur's agriculture is so intensive that it provides an important revenue even though it occupies only a small percentage of the total surface area. Industrial activities there are by no means negligible – all the more so because businesses that have recently set up there are often 'high-tech'. Service industries intended for business enterprises willingly open in places where it is easy to attract executives, and where air transport provision is excellent.

In certain instances, the dominant feature is the variety of activities: agricultural land, as elsewhere given over to a monoculture or a small number of allied ventures, is very productive, but hardly counts beside frequently high-performance and sophisticated industries and extremely diversified tertiary enterprises. One must then speak of *regions of complex economy*,[3] in order to contrast them with simpler regions the profiles of which are sufficiently defined by their dominant specialization.

Economic analysis also reveals the pertinence of the effects of polarization: areas of commercial attraction in a town constitute the envelope for the trade that give it its livelihood or represent one of its fundamental functions. The services provided in the heart of this area allow it to steer and organize. Not all urban centres carry the same weight: their zones of influence form series which fit into one another according to the hierarchic classification of the towns concerned. One level was of interest to geographers in the late 1930s: *urban-based regions* dominated by a veritable *metropolis*, that is, by a centre of direction, coordination and impetus that dominates an entire territory. The regional metropolis may be recognized by its banks, the presence of headquarters or other offices of major companies, an independent press, a cultural life enlivened by theatres, an opera house, orchestras and the intellectual stimulus provided by the presence of one or more universities. A certain administrative autonomy often completes the picture.

Regional differentiation of inhabited territory is naturally built around production. The other side of economic life – consumption – has far less significance in regional analysis: within a single area, habits and purchasing levels are often very uniform. They hardly vary except with the hierarchic level of the towns: customers with the highest incomes frequently reside in capitals and major cities.

Study of social realities proves disappointing for establishing regional divisions. The categories are generally too interwoven for certain areas to be characterized by the predominance of one profession or one class. There are exceptions: the proportion of farmers in rural areas is often overwhelming, but they are never the only people to live there. Variety of status and occupations is the first characteristic of towns, though that does not prevent fine shadings. Some centres are almost exclusively oriented towards industry, whereas others have their economic basis in services. The most highly qualified and best-paid jobs are more numerous in large towns. Geographers cannot remain unaware of the tensions that result from these uneven distributions but these are not sufficiently clear-cut to serve as a guide for the divisions they are seeking to establish.

It is often a completely different matter with regard to cultural realities: certain territories contrast with others because of the origin of their inhabitants, language or dialect, religion or lifestyle. But the ensembles resulting from these splits vary greatly in size, and their contours sometimes lack regularity. Whereas in a trading economy, areas that specialize in differing products are of comparable size and cover the whole national area, cultural criteria sometimes lead to the distinguishing of tiny cells buttressed by their traditions in the midst of a fairly homogenous ocean where everyone speaks the same language, subscribes to the same ideals, practises the same religion and espouses the same ideologies. Because of the irregularity of its bases, the *cultural region* has thus attracted less attention than the economic region. There is one exception, however: among the features of material culture, some contribute outstandingly to modelling the landscape. This is the case of agrarian structures. Between 1930 and 1960, these gave rise to remarkable research. Thanks to this, much knowledge was gained of the distribution of field systems and the forms and types of enclosure. From this work, it emerged that the morphological features of the landscapes were closely interrelated, which explains their surprising stability over long periods – and their violent change in periods of fluidity.

It is known today that divisions relating to a particular characteristic are often contained within limits that differ greatly from those established for others. This is what makes transition from an analysis of isolated features to a synthetic view so difficult. At the beginning of the century, geographers argued differently; they were convinced that physical factors had so much influence on the life of groups that natural regions constituted the framework within which human realities were also moulded. That is the definition they gave to *geographical regions*. In the period between the wars, as their positions evolved, it was realized that relief, climate, vegetation, agricultural life, trade and cultural affairs were generally contained in areas that were not superimposed. The imperatives of relationships ended in systematic clashes between natural regions and areas of economic solidarity; in the conditions of traditional life, when transport proved difficult, trading could flourish only over short distances: commerce was born of the complementary nature between rural areas or the contrast in zones of altitude, and the urban centres which carried out the transactions located on the boundaries of the natural ensembles.

A large number of natural, cultural and economic factors may be contained within the same area. The term geographical region may be employed here: this is a category that is all the more interesting because it is exceptional. As Pierre Birot emphasized strongly, in 1949, describing the case of Portugal: 'Thus the Minho is the realization of one of the most perfect types of geographical region imaginable. The alternation of the seasons, the differing characteristics of a heterogeneous soil compete to determine the components of a harmonious landscape and agricultural system'.[4] In order to define the rural civilization of these lands, he resorts to the term of agrarian 'climax'. This *pays* has not escaped external influences and has evolved, 'but what has guaranteed the permanence and unity of the geographical region is that these transformations have taken place simultaneously over the whole of the territory and without any revolutionary upheavals, new plants and techniques having been merely assimilated by the old agrarian system'.[5] Time passes without weakening the features peculiar to the whole of the territory. Birot continues:

> The Minho example also shows us that a true geographical region is something unique [. . .] the combination of all those characteristics as realized in the Minho could not be repeated twice on the earth's surface. That is why, although general physical and human geographies are natural sciences, since they classify types reproduced from a certain number of interchangeable models, regional geography is an art that associates itself with true individualities. It is not unlike the feeling of the biographer for his hero, a love for something whose like will not be seen again.[6]

*The region as a life-world**

Geographers have recognized other types of grouping, but somewhat differing in nature from the units listed so far. Local characteristics noted and taken into account when there is a change of scale were hitherto of an objective kind; this was true of relief, climate, production or the shape of fields. But we find very early mention of *pays*[7] and historical regions in works relating to natural regions.

The first term is applied to regions that are fairly small in size. Their particular feature is to form a background recognized by the people who live there and by those who live round about. A simple natural region is perceived only by those investigating, whereas one that is also a *pays* is the subject of a representation shared by an overall group: geographers are not the only ones to carry out divisions in surface areas. A division changes its nature when it is taken over by a collective consciousness.

The historical region refers to occurrences of the same nature. Analysis sometimes reveals groupings of which everyone is aware, which are pinpointed by the names that are attached to them, to which people show a particular loyalty, and which do not correspond to any objective division that can be discerned today. Coincidence with natural units is not borne out – historic regions often gather together varying and complementary milieux. Cultural homogeneity is not essential – a good number of historic regions are inhabited by people who do not speak the same dialect or even the same language, and who do not

* See translator's note on p. 22.

attend the same churches. These regions are sometimes dominated by a town, but there is no absolute rule, and the city that symbolizes their unity has sometimes greatly declined nowadays.

Around what were historic regions built? Around a consciousness of a common destiny. This consciousness may be rooted in a distant past the early stages of which are forgotten: a certain number of historic regions which remain alive in France are the heirs of the Gallic tribal groupings established in the third century BC – for example, the Auvergne, Franche-Comté and Limousin. Elsewhere, in Normandy, the original grouping was due to Diocletian's administrative reforms in the Late Roman Empire, and was reinforced by the arrival of the Vikings and the gift made to Rollon* by the king of France. Burgundy recalls a series of territorial constructions connected, in very early days, with one of the waves of Germanic invasions.

Historic regions, therefore, have not all emerged from the same type of event and are not embedded equally deeply in the past. But what they have in common is that they have given rise to a collective consciousness: an area of land exists the identity of which is strongly perceived by a collective group; the people who live there share certain features. From that starting point, the feeling of belonging is perpetuated from one generation to the next: the geographic unit is a mental construct.

The typologies resulting from the expansion of the regional concept in the first half of this century are somewhat paradoxical: they are built on heterogeneous criteria and do not all refer to the same levels of reality. But their apparent imperfection leads to their being passed over. The aim of regional geography is not to classify and establish typologies. It is to show how varied the organization of space can be: we must analyse the mechanisms at work, and see how and why, depending on periods of time, they have resulted in variable considerations.

Further Reading

Buttimer, A., *Society and Milieu in the French Geographic Tradition*, Chicago: Rand McNally, 1971.

Cloke, P., Philo, C. and Sadler, D., *Approaching Human Geography: An Introduction to Contemporary Theoretical Debates*, London: Paul Chapman, 1991, pp. 3–13.

Dickinson, R. E., *The Regional Concept: The Anglo-American Leaders*, London: Routledge and Kegan Paul, 1976, 408 pp.

Fisher, C. A., Wither regional geography?, *Geography*, vol. 55, 1970, pp. 373–89.

* A tenth-century Norman chief, who was ceded land by Charles III in 911 that assumed the name of *Normandie*.

Herbertson, A. J., The higher units: a geographical essay, *Scientia*, vol. 14, 1913. Reprinted with commentary articles in *Geography*, vol. 50, no. 229, pp. 313–72.

Johnston, R. J., Hauer, J. and Hoekveld, G. A. (eds), *Regional Geography: Current Developments and Future Prospects*, London: Routledge, 1990, 216 pp.

Kimble, G. H. T., The inadequacy of the regional concept, in Stamp, L. D. and Wooldridge, S. W. (eds), *London Essays in Geography*, Longman, 1951, pp. 151–74.

Lewis, M. W. and Wigan, K. E., *The Myth of Continents: A Critique of Metageography*, Los Angeles and London: University of California Press, 1997, 344 pp.

McDonald, J. R., The Region: Its Conception, Design and Limitations, *Annals of the Association of American Geographers*, vol. 56, 1966, pp. 516–28.

Patterson, J. H., Writing regional geography: problems and progress in the Anglo-American realm, *Progress in Geography*, vol. 6, 1974, pp. 1–26.

Pudup, M. B., Arguments within regional geography, *Progress in Human Geography*, vol. 12, 1988, pp. 369–90.

Sauer, C. O., *The Morphology of Landscape*, University of California Publications in Geography, no. 2, 1925. Reprinted in Agnew, J., Livingstone, D. N. and Rogers, A. (eds), *Human Geography: An Essential Anthology*, Oxford: Blackwell, 1996, pp. 296–315.

Whittlesey, D. et al., The regional concept and the regional method, in James, P. E. and Jones, C. F. (eds), *American Geography: Inventory and Prospects*, Syracuse University Press, 1954, pp. 19–68.

3
Methods of Regionalization

Regional investigation takes place at different levels, that of the commonplace experience of all and sundry, or that of abstract research into causes, correlations and influences. Whatever its aims may be, it needs prerequisites. It is not enough, at any one point, to note an unprecedented phenomenon to raise a regional problem: one must know whether the phenomenon exists there only or whether it is apparent over a certain expanse and, if so, how far. Definition of areas presenting this or that characteristic is essential to determining the range of observation and researching its causes. All regional approaches rest on the ability to divide an area according to the variations of any of its features, or a whole group of them. An understanding of how data are organized is crucial to appreciating the overall logic of the operations that lead to grasping reality in all its complexity. Therein lies the empirical foundation of regional geography.

The boundaries revealed by cartographic or statistical procedures are not all on the same model. Some correspond to a clear discontinuity, whereas in other instances, passage from one region to another is by gradual transitions, or by a mosaic in which the basic areas show contrasting characteristics. In regional geography, the nature of the frontier zones is as significant as that of the areas where this or that feature (or combination of features) asserts itself.

3.1 From Land Survey to the Map: Genealogy of the Methods of Regionalization

The individual survey

The initial phase of any regional work consists of proceeding to divide, in order to demarcate within a territorial ensemble, sub-areas the elements of which resemble one another more than they resemble those of neighbouring units.

There are several ways of undertaking these operations. Some are purely intuitive. Others are the result of the systematic observation of the terrain; they may rely to a fair extent on a brief inspection, on individual assessment, but they are proven by the scientific training received by those who practise them. A geologist, a naturalist or a geographer picks out elements which elude those who lack the same knowledge: they know how to recognize, in reality, associations or configurations that have significance for anyone seeking to understand spatial order.

Direct analysis soon comes up against limitations: it is impossible for the same individual to follow every road and path, to carry out every search and establish every transect essential to grasping in equally minute detail each fraction of the ground being studied. If the approaches relied on the work of one person alone, a precise regionalization would be possible only in very confined areas. On a smaller scale, one would never have anything but indications that were too limited to be trustworthy. The divisions they suggest to the well-trained observer are often interesting, but they are merely hypotheses that need to be checked by means of a more complete analysis.

Regionalization requires investigations that are too demanding to be carried out by individual researchers. The latter, in order to verify that their intuitions are well founded, need to have the backing of colleagues who complement their work by extending it to zones they cannot survey themselves, or to make use of data supplied by a regular system of observations.

The systematic survey

When conducting a rigorous exercise in dividing up an area, in order to have sufficient information it is necessary to make use, following standardized procedures, of data obtained by a number of observers. Each submits to the same protocols so that the overall results form a coherent whole. As far as possible, subjective appraisals are replaced by measurements: one no longer notes that the atmosphere is dry, humid or stifling, but records temperatures, wind speed and direction and hygrometric degrees.

The work of systematic surveying allows one to go beyond what is revealed by the landscape: the hurried traveller sees only relief features, vegetation, and the external aspects of settlement (architecture, grouping or scattering of houses, lands arranged in huge bare spaces or divided into enclosed fields) and population (physical type, clothing, churches, festivals, fairs, markets, etc.). The moment there is access to counts taken by administrative units, simple calculations can be made. Density (relation between the number of inhabitants and the surface they occupy) raises very precisely the problem of the relationship of groups with the environment in which they live, on which they depend wholly or partly, and on which they exercise ecological pressures. As early as the end of the nineteenth century, France gave pride of place to a comparison of densities. That permitted the posing of fundamental questions on the relationship of societies to space.

Generalization

For anyone wanting to divide an area of land into sub-units, the classic method relies on the mapping of data. In many fields there are precise records covering a large number of points. They are often close enough to let one interpolate, that is to say, suppose that between adjacent places the values vary in a continuous manner. One thus ends up with a satisfactory map of the phenomenon and ensembles begin to appear: this is visual generalization.

In many other cases, the data define the average state of a basic area rather than a precise place. This was often so for human, social or economic facts, where one is dependent on censuses and other data gathered by administrations. The finest accessible unit is generally the district or commune. The work of regionalization rests on cartographic analysis of ranges of values characterizing these primary units.

3.2 The Map as Instrument of Regionalization

Highlighting homogeneous regions on the basis of primary units defined by a single characteristic

To understand how data are regionalized, it is as well to start with an instance where each point of an area is defined by a unique qualitative characteristic. When it is present it is given the value 1, and its absence is denoted by 0.

Let us transfer the data available to us for an ensemble of districts to a base map on which their boundaries are shown (figure 3.1), and observe the distribution that emerges. Three instances may present themselves. In the first (figure 3.1a), the units where the characteristic is present form a compact whole to the north of the area studied, and are totally lacking in the south. In the second (figure 3.1b), there is an inextricable mixture of districts where the characteristic is present or absent. In the third (figure 3.1c), there is a first ensemble (A) to the north where all the communes show the characteristic, a second (B)

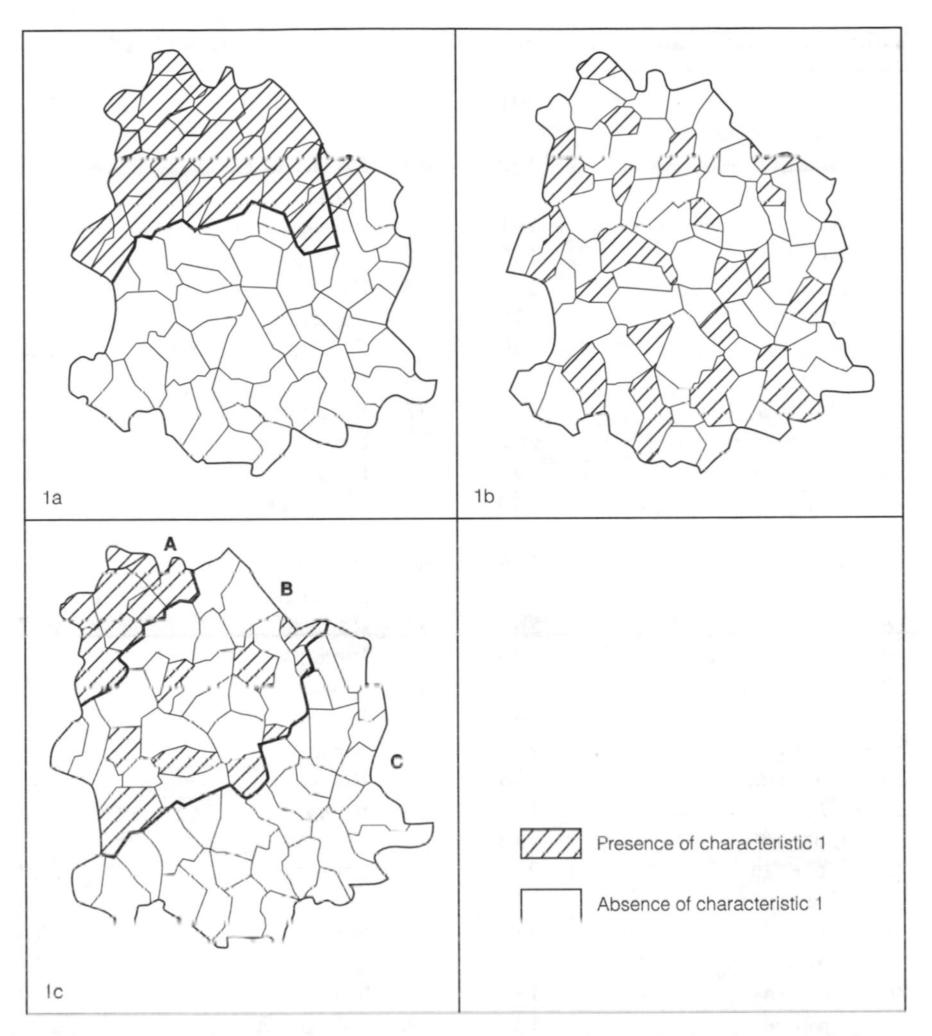

Figure 3.1 Regionalization of a distribution with a single characteristic

in the centre where there is a mixture, and a third (C) to the south where the mapped feature is absent.

In the first case, cartography shows that regionalization is possible and allows a distinction to be made between two compact homogenous wholes. In the second, the data reveal no spatial order: classification based on the selected variable leads to no regional division. For that to be possible, one must ascertain that the elements possessing the same characteristic are contiguous.

The third, more frequent, case discloses a more complex situation: regionalization is possible, in the sense that homogeneous wholes can be seen taking shape in A and C. But it is zone B that raises problems: it is a mosaic, some pieces of which are similar to A and the others to B. There is therefore a fringe of indecision between the two wholes

Table 3.1 Birth rates in Italy in 1960 (rate per 10,000 inhabitants)

Province	Rate	Province	Rate
1 – Agrigento	201	47 – Messina	189
2 – Alessandria	113	48 – Milano	160
3 – Ancona	152	49 – Modena	152
4 – Aosta	144	50 – Napoli	269
5 – Aquila	161	51 – Novara	138
6 – Arezzo	142	52 – Nuoro	219
7 – Ascoli Piceno.	172	53 – Padova	195
8 – Asti	110	54 – Palermo	232
9 – Avellino	199	55 – Parma	126
10 – Bari	249	56 – Pavia	117
11 – Belluno	139	57 – Perugia	152
12 – Benevento	200	58 – Pesaro	155
13 – Bergamo	191	59 – Pescara	170
14 – Bologna	130	60 – Piacenza	126
15 – Bolzano	215	61 – Pisa	136
16 – Brescia	186	62 – Pistoïa	135
17 – Brindisi	240	63 – Potenza	227
18 – Cagliari	247	64 – Ragusa	207
19 – Caltanisetta	250	65 – Ravenna	149
20 – Campobasso	171	66 – Reggio Cal..	228
21 – Caserta	248	67 – Reggio nell'E..	122
22 – Catania	248	68 – Rieti	147
23 – Catanzaro	258	69 – Roma	202
24 – Chieti	154	70 – Rovigo	182
25 – Como	152	71 – Salerno	235
26 – Cozenza	235	72 – Sassari	218
27 – Cremona	146	73 – Savona	128
28 – Cuneo	128	74 – Siena	116
29 – Enna	209	75 – Siracusa	212
30 – Ferrara	151	76 – Sondrio	199
31 – Firenze	136	77 – La Spezia	130
32 – Foggia	242	78 – Taranto	236
33 – Forli	168	79 – Teramo	170
34 – Frosinone	182	80 – Terni	137
35 – Genova	122	81 – Torino	146
36 – Gorizia	144	82 – Trapani	205
37 – Grosseto	137	83 – Trento	175
38 – Imperia	139	84 – Treviso	183
39 – Latina	218	85 – Trieste	102
40 – Lecce	217	86 – Udine	135
41 – Livorno	141	87 – Varese	172
42 – Lucca	149	88 – Venezia	194
43 – Macerata	155	89 – Vercelli	117
44 – Mantova	149	90 – Verona	177
45 – Massa Car.	144	91 – Vicenza	194
46 – Matera	231	92 – Viterbe	157

that we have selected. One could equally say that between the two compact masses, A and B, there exists a transitional zone which may be differentiated, and where the distributions observed form a mosaic.

What happens when the characteristics are not discrete sizes but continuous variables? For example, we may consider a series of data (table 3.1) on birth rates in the Italian regions in 1961. The rates vary from 110 births per 10,000 inhabitants in the province of Turin to 269 in that of Naples. It may be considered that provinces the rates of which are included in sections of a magnitude of 50 are close: the first map (figure 3.2a) was thus constructed by attributing a single value to all the zones where the birth rate varies between 100 and 149 per 10,000; 150 and 199; 200 and 249; and 250 and 299. The contrast between the North-West, Central and Southern Italy may be clearly seen. But the cartographic representation is not perfect: the shading for the zones where birth rate is equal to or above 250 per 10,000 is used only three times. It would seem wiser, if one wishes to obtain a more meaningful map, to ensure that each shading value corresponds to a single number of provinces. To achieve this, it is enough to construct the histogram of cumulative frequencies (figure 3.2b), and retain on it sections corresponding to the 25 per cent with the lowest birth rate, the 25 per cent following, etc. The map obtained (figure 3.2c) is more expressive than the first. It may however be faulted for not taking account of the distinctive features of the recorded distribution. These appear on the diagram of frequencies (figure 3.2d): the curve is bimodal, with a first ensemble of regions with a high birth rate, and a second where the rate is average or low. When one is combining into the same class the individual cases that have the most similarity, must not the first concern be to indicate the dichotomy that the double maximum reveals? This leads us to distinguish between wholes that the diagram of frequencies exposes, which are then subdivided; the map (figure 3.2e) expresses this: the specific nature of the Mezzogiorno is fully revealed, whereas its boundaries with the zones of central Italy had previously lacked clarity.

Highlighting homogeneous regions when the primary units are defined by two characteristics

What happens when each point, or each primary area, is defined by two characteristics (1) and (2)? To simplify, we will limit this to the case where only the two values 0 and 1 are possible.

Mapping demonstrates several types of possible configurations. First, there is the case where (1) and (2) are scattered at random: the data cannot be regionalized (figure 3.3a). In figure 3.3b, the primary units

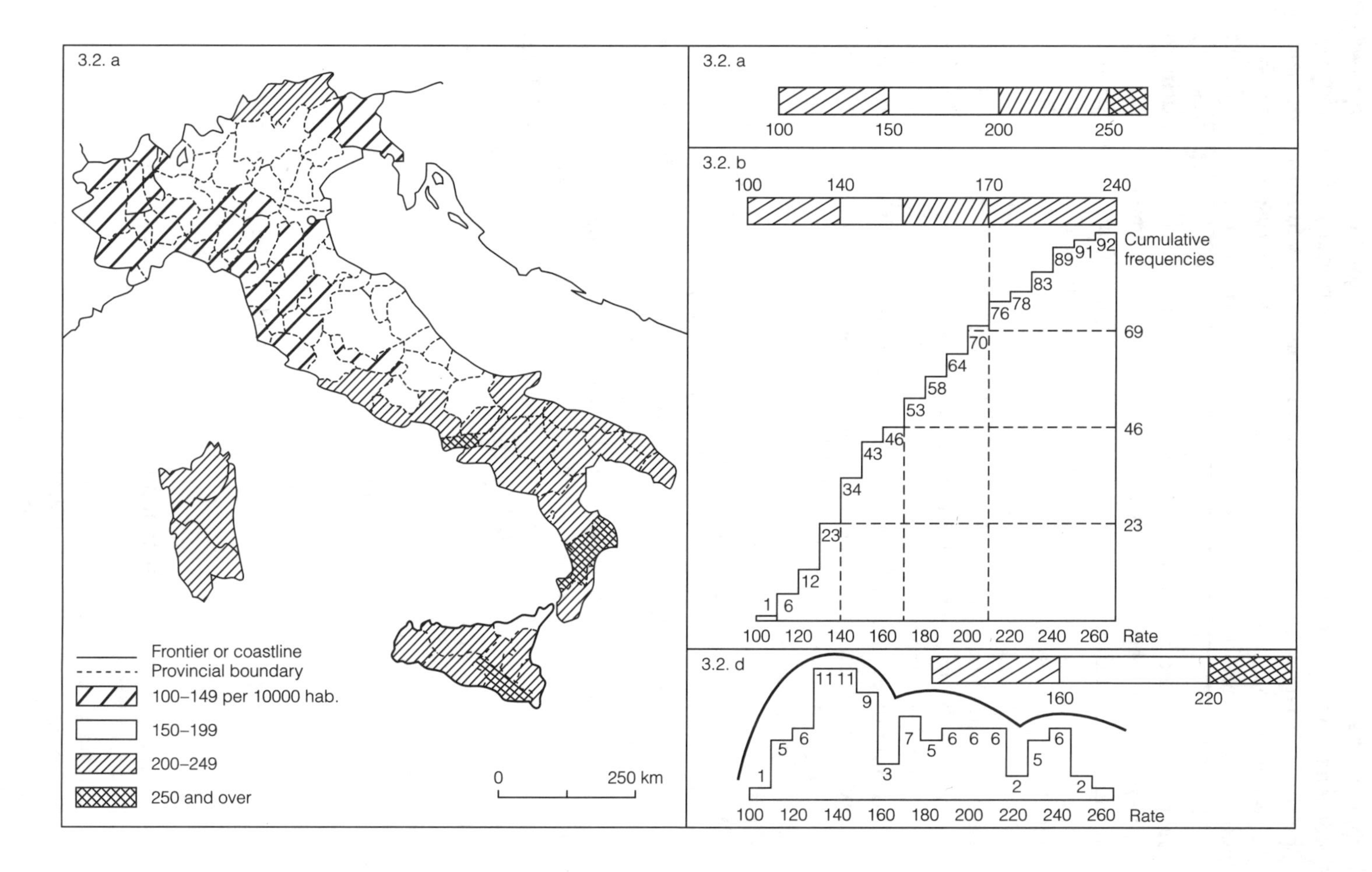

3.2. a
Frontier or coastline
Provincial boundary
100–149 per 10000 hab.
150–199
200–249
250 and over
0
250 km
3.2. a
100
150
200
250
3.2. b
100
140
170
240
Cumulative frequencies
1
6
12
23
34
43
46
53
58
64
70
76
78
83
89
91
92
69
46
23
100 120 140 160 180 200 220 240 260 Rate
3.2. d
160
220
1
5
6
11 11
9
3
7
5
6 6 6
2
5
6
2
100 120 140 160 180 200 220 240 260 Rate

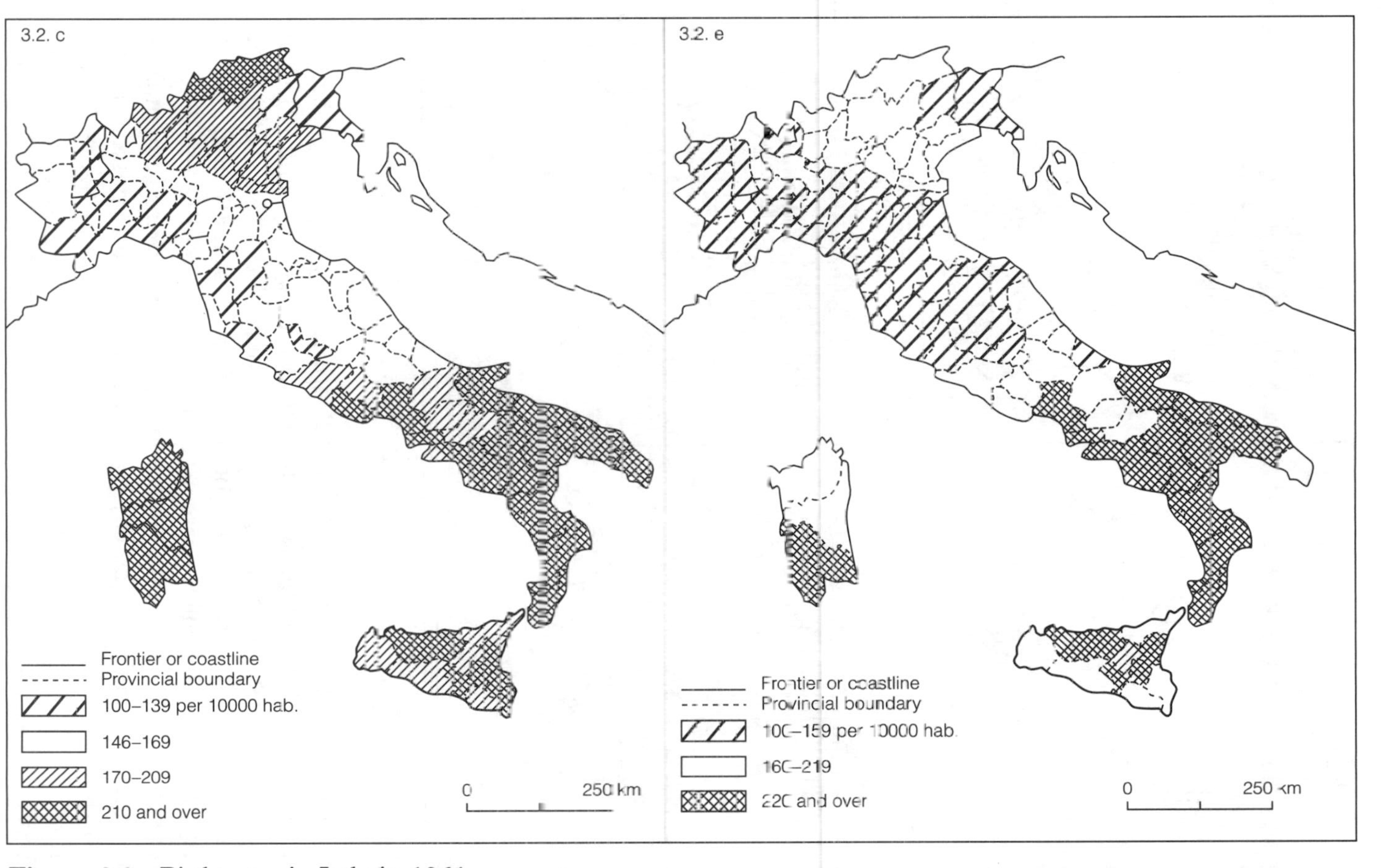

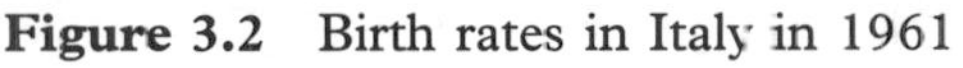

Figure 3.2 Birth rates in Italy in 1961

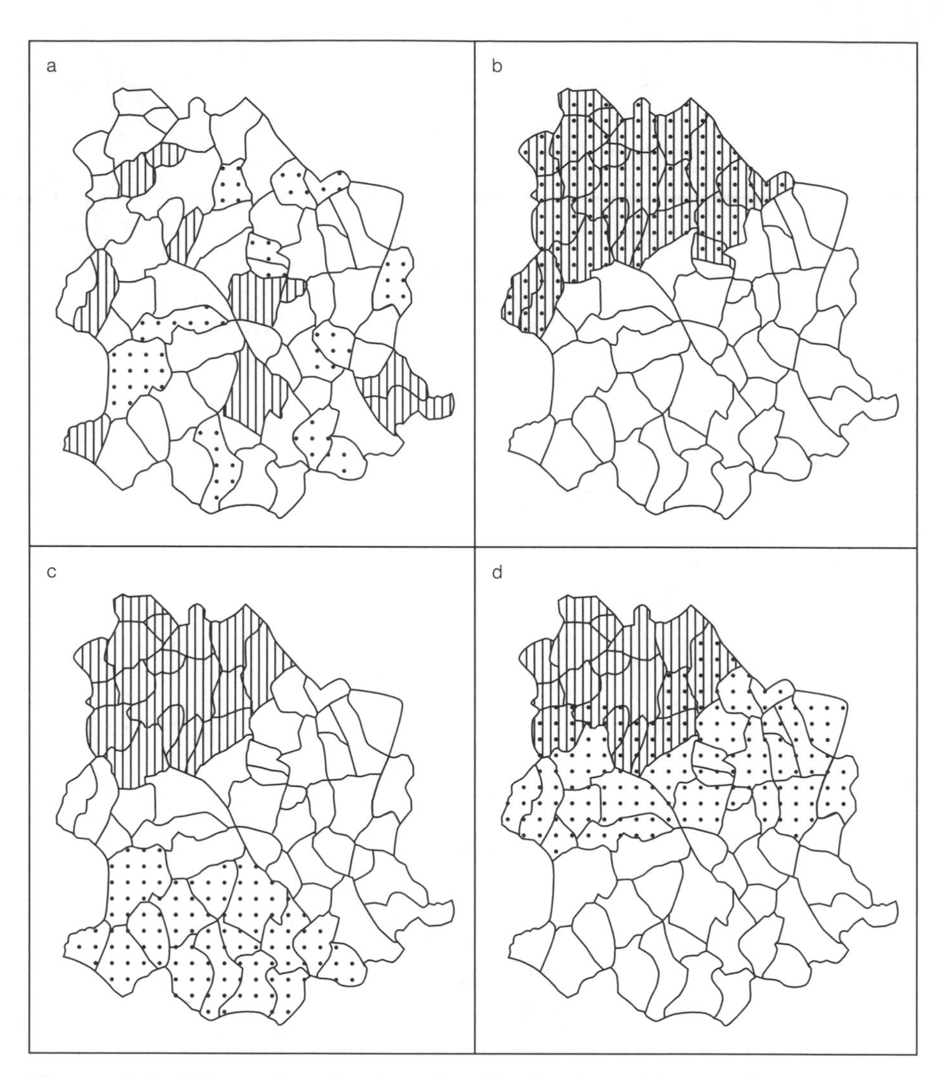

Figure 3.3 The regionalization of a distribution with two characteristics

possessing characteristics (1) and (2) are exactly superimposed: two territorial wholes may be seen, the one comprising (1) and (2) in the north, and the one containing neither (1) nor (2) in the rest of the area under study. But there may also be divisions such as those in figure 3.3c: here, the districts which possess (1) and are situated in the north do not possess (2). Those with this second characteristic form a homogeneous ensemble in the south. The remainder in the centre possess neither.

In many instances, distributions like that shown in figure 3.3d are more apt to be observed: the distributions of (1) and (2), as in the preceding case, form continuous blocs. They can thus both be regionalized, but here we have: (i) an east-west band where (1) and (2) coexist; (ii) towards the north a zone that extends it where only (1) appears; (iii) in the centre an area that experiences only (2); (iv) a vast expanse to the south where neither (1) nor (2) is present. One therefore sees four regions take shape on the basis of a distribution with two characteristics.

The case of distributions with n *characteristics*

Let us now suppose that, for each of these primary areas, we process *n* number of data. How do we proceed in the face of such wealth?

The simplest way is to establish at the same scale a map relating to each of the available characteristics. Certain data are distributed randomly or formed from a regular mosaic: they cannot be regionalized, whereas others can be. One then compares the limits characterizing each distribution by superimposing them. Several situations are feasible:

(1) All the outlines are similar, even if they are not precisely superimposed. The map obtained by tracing all the boundaries of the primary regions throws up nuclei of an often appreciable area where a whole set of characteristics are present simultaneously and, round about, areas where the boundaries follow and intersect one another so that it is not possible to speak of homogeneity there. There are thus homogeneous regions separated by transitional zones (figure 3.4a).

(2) The outlines do not all resemble one another, but they can be grouped in two, three or several families. For each the same configuration is observable as in the preceding case: homogeneous nuclei and transitional zones. Several regionalizations are therefore possible, but they cannot be combined to give a single synthetic configuration (figure 3.4b).

Different methods of procedure are often used. Instead of embarking on the lengthy work of mapping each of the available data, one builds them into a synthesized resumé. In climatology, for instance, empiric formulae are thus available that integrate average temperature, its effectiveness on vegetation measured in terms of evapotranspiration, the length of the vegetative season and rainfall. Each point, or each primary unit, can therefore be allocated a class. When the result is transferred to a map, one sees either the appearance of homogenous ensembles, mosaics, or confused divisions that show regionalization to be impossible.

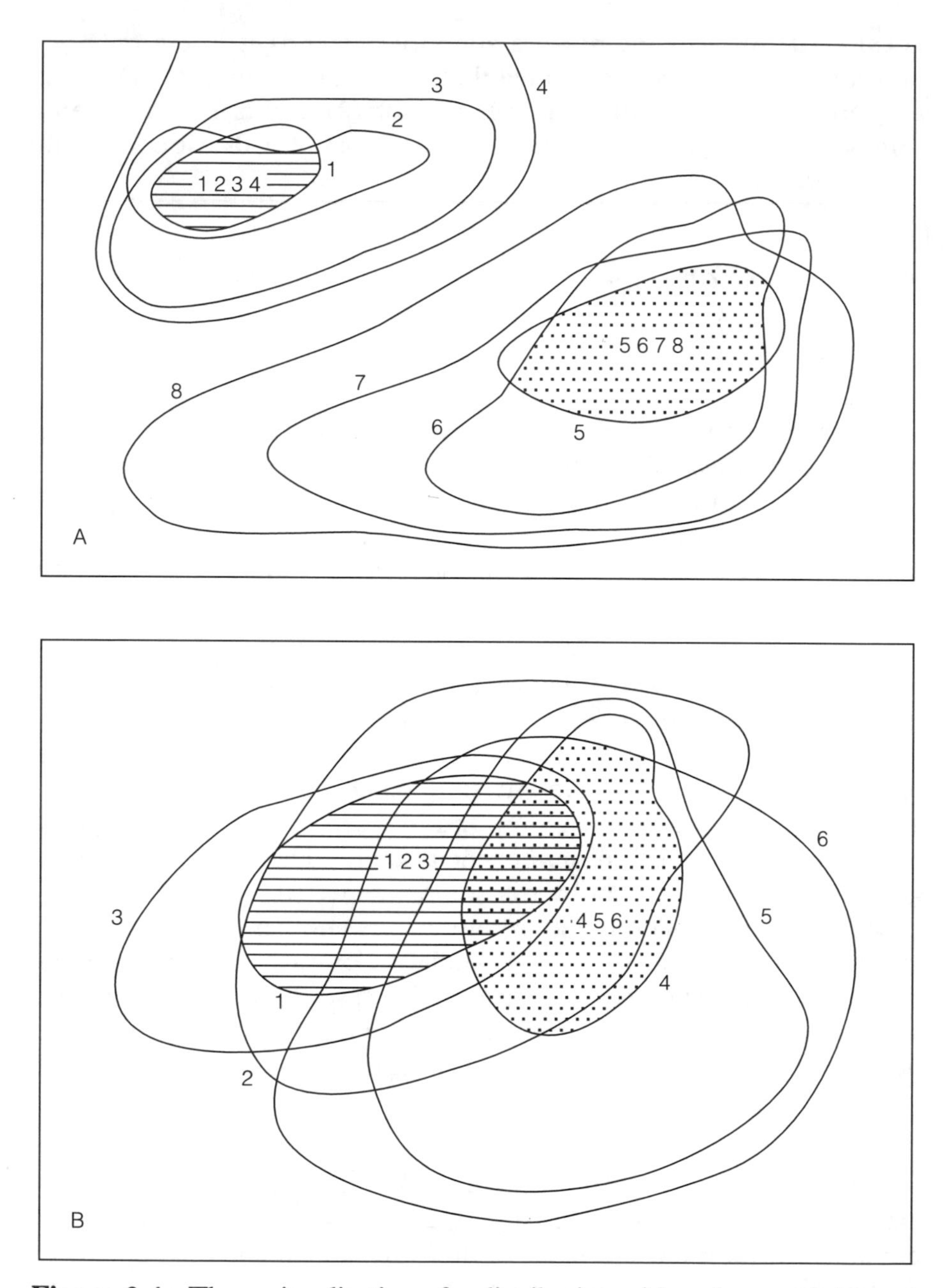

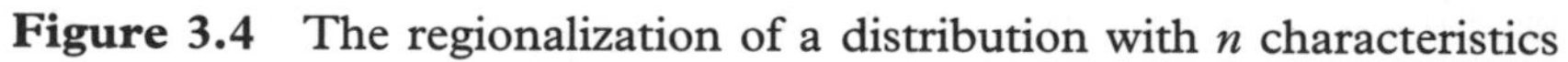
Figure 3.4 The regionalization of a distribution with *n* characteristics

When one starts from a basis of raw data to construct, by grouping, a display of units that present the same features, one is said to have followed an ascending regionalization procedure. When the data are combined in such a way as to replace them with a synthetic index, the procedure is termed descending. This type of operation is most frequently used in climatology.

Defining polarized regions

The data mapped relate to a point or to a primary area of analysis. The regions which are defined are homogeneous because what they have in common is to possess the same characteristic(s), or to have it (or them) allocated identical or similar values. The flows that develop in a given area are equally of value: do they not reveal the exchanges that unite its subsets and help an understanding of their functioning? As economies have become more open and the mobility of people, goods and information has grown, geographers have become more attentive to these aspects of reality, which for long they had hesitated to take into account because it meant dealing with elements that could not be read directly from the landscape.

Points or primary areas may have in common the maintenance of the bulk of their relations with the same centre. The ensemble thus formed possesses a special kind of homogeneity: the one that is born of a permanent dialogue with a common focal point – a dialogue that means a complementary nature but, also very often, dependence. This particular type of grouping is known as a *polarized* or *nodal region*.

Statistics indicating flows are often provided for primary areas rather than for individual points. To make use of them cartographically, one generally settles for portraying the flows between geometric centres, centres of gravity or the chief towns of districts. This has been done in figure 3.5. We can see in figure 3.5a that the areas are arranged around three centres, A, B and C, without the flows overlapping. The criterion of contiguity is respected and regionalization presents no difficulty. In figure 3.5b, certain links overlap, which leads one to make a distinction between nuclei with firm contours, and intermediate zones where there is ambiguity. Lastly, in figure 3.5c, centre B and C cease to appear independent: they are both turned towards pole A, which is more powerful. Hierarchization appears in the levels of the poles, while the areas they dominate are dovetailed.

The limitations of cartographic techniques

Until the beginning of the 1960s geographers regionalized the data available to them almost solely by using cartographic techniques. These offered many advantages: they showed up groupings at a first glance and immediately differentiated between regionalizable characteristics and those which were not. When numerous sets had to be dealt with, superimposing analytical maps highlighted the complex structure of the area, central homogeneous zones and the more varied aureolae surrounding them. Mapping of flows revealed the degree of polarization.

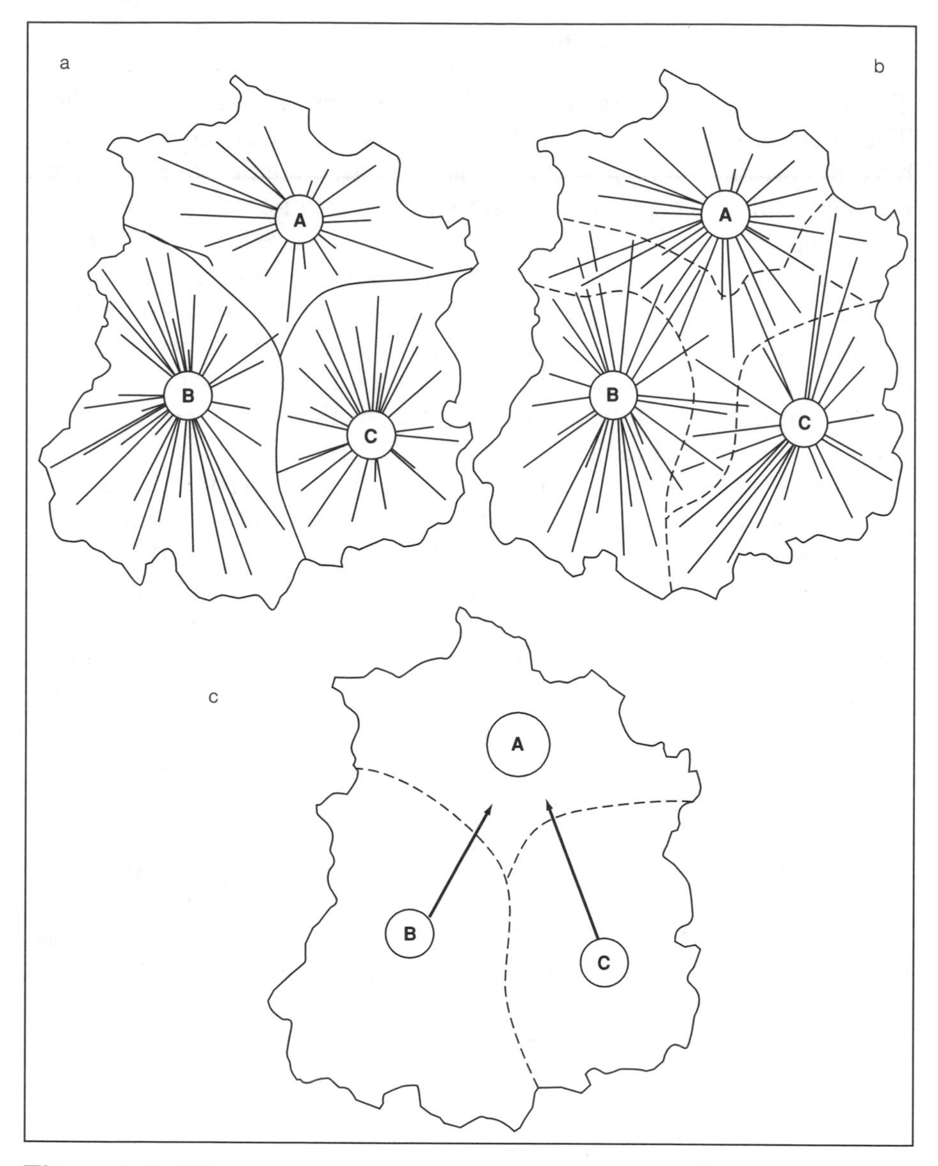

Figure 3.5 The determination of polarized regions

But this method was perfectly satisfactory only when one was content to map a single characteristic. As soon as several had to be dealt with, it became evident that the demarcations suggested did not match. Regionalization then rested on a rather subjective interpretation of the results.

Spatial statistics were insufficiently developed. Sociologists and economists chiefly handled aspatial data for which they had at their disposal treatment procedures that were well proven. Historians employed systematic means to approach chronological sets. In the past thirty years, the study of chronological sets has made great strides.

3.3 Regionalization as a Special Case of Classification Methods

The turning point at the beginning of the 1960s

In 1961, two of the geographers who launched the 'new' geography movement published an article that at the time passed unnoticed: J. D. Nystuen and M. F. Dacey[1] in fact exploited graph theory to systematize the procedure previously used to explore nodal structures. The various points of the area are linked by vectors. The matrix to which their values are transferred allows classifications hitherto established on cartographic bases to be effected by calculation. This work gained in objectivity. It is facilitated when data are too numerous to be easily transferred to the same diagram.

William Bunge[2] brought the argument into general use: in *Theoretical Geography*, published in 1962, he emphasized that, in fact, to regionalize was no different from classification in any discipline, but with the addition of a spatial term that is lacking elsewhere: in order that similar elements may be grouped, they must be contiguous. The application of modern statistical procedures was exploited several years later by Peter Haggett[3] in *Locational Analysis in Modern Geography*. The work of Grigg,[4] exactly contemporaneous, went in the same direction.

Upholders of the new geography at the time blamed their precursors for having laid such emphasis on the specifics of their discipline that they had ended by forgetting the rules of method that were valid for any scientific approach, and for attributing an 'exceptionalist' character to their discipline. By showing that regionalization was only a special instance of methods of classification, the Young Turks of the 1960s taught that there is much to be gained by drawing on common experiences. Regional analysis ceased to rely on intuition, and frequently impressionistic cartography, the moment it took its example from what was being done elsewhere.

The progress achieved was important: it revealed the logical bases on which the procedures until then had rested, but it did not completely change their results: the divisions that emerged bore a very close resemblance to those provided by mapping. The same approach could result in varied classifications: a territory could be divided into 1, 2, 3, 4, 5 or 6 classes. What could henceforth be measured was the content to be given to the idea of homogeneity, and the limits that could be imposed on it.

The logic of regionalization

Let us take the simplest instance: the space to be studied is made up of a collection of primary areas each characterized by the presence or absence of a characteristic, which can be recorded by the figures 1 and 0. The whole of the data available thus forms a reduced vector in which peaks take the values 1 or 0. To classify is to put together every 0 and every 1. Transfer to a map allows one to see whether or not there is contiguity, as we noted above.

Most of the data are measurable but, as the units used change from set to set, intervals of variation are very different. Luckily it is always possible to standardize a statistical set, that is, to replace the original unit of measurement with a range whose variation is included between −1 and +1 or between 0 and 1. The case we have just analysed is the borderline case in which the variable can have only two values, 1 and 0 or 1 and −1. Generally speaking, it can have any value within the interval of variation. A primary area belongs completely to the region when its characteristic equals 1, but if it equals 0.9, 0.8 or 0.7, it continues to be part of it, though in a decreasingly clear-cut manner. The theory of fuzzy subsets shows how inclusions evolve when the value of the accepted characteristic gradually declines. The work of Tran Qui[5] opened up the way in this field during the 1970s.

However, procedures based on the theory of fuzzy subsets are not generally brought into play when regionalizing data: they underline the general logic of the operations being carried out, but do not supply the most operational means.

Ascending hierarchic classifications

Let us suppose an ensemble of *m* primary areas each of which is defined by values assumed there by *n* characteristics. The information contained in the set is presented as a matrix where *m* primary areas are in rows and the *n* characteristics are in columns. How do we deal with the information on this matrix so as to allow regionalization?[6] After standardizing the data, one calculates the distance, in the *n* dimensional area of observations, separating the characteristics corresponding to each *x,y* pair of areas, and starts afresh for all the possible pairs. The general formula which gives these distances is:

$$D = \sqrt{\sum_{i=1}^{n} (x_i - y_i)^2}$$

in which the characteristic i is included between 1 and n.

It is then possible to arrange all the information contained in the set of data relating to the primary areas by constructing a squared matrix of *m* rows and columns to which all the distances obtained are transferred. Thus a squared matrix of the distances between the primary areas replaces the initial rectangular matrix of the areas and the characteristics.

If we have six units of observation, A, B, C, D, E and F, we obtain the matrix 3.1a. It is a symmetrical matrix since the distances figure twice, on either side of the diagonal. The two areas the characteristics of which are closest are D and F, since the distance separating them is only 1 (in italics on the matrix). Regrouping them, one arrives at a matrix that has no more than 5 rows and columns (matrix 3.1b). Points A and C (distance 2, in matrix 3.1c) are then the ones that fuse. B and E (distance 3, in matrix 3.1d) may be regrouped. The regrouping shown in matrix 3.1e implies a larger jump since A, B, C and E are joined starting from the value 6.6. The ensemble can form a single class only if one accepts that the distance between the subgroups is 10.25.

Matrix 3.1a

	A	B	C	D	E	F
A	0	6	2	14	9	15
B	6	0	4	8	3	9
C	2	4	0	12	7	13
D	14	8	12	0	5	*1*
E	9	3	7	5	0	6
F	15	9	13	*1*	6	0

Matrix 3.1b

	A	B	C	DF	E
A	0	6	*2*	14,5	9
B	6	0	4	8,5	3
C	*2*	4	0	12,5	7
DF	14,5	8,5	12,5	0	6
E	9	3	7	5,5	0

Matrix 3.1c

	AC	B	DF	E
AC	0	5	13,5	8
B	5	0	8,5	*3*
DF	13,5	7	0	5,5
E	8	*3*	5,5	0

Matrix 3.1d

	AC	BD	FE
AC	0	6,5	13,5
BE	6,5	0	7
DF	13,5	8,5	0

Matrix 3.1e

	ACBE	DF
ACBE	0	10,25
DF	10,25	0

Figure 3.6a expresses the results provided by the matrix in the form of a hierarchic cluster analysis in which the thresholds of fusion are clearly in evidence. The diagrammatic maps (figure 3.6b) show the various configurations according to the grouping threshold used.

The method just mentioned is that of an ascending hierarchic classification. There are different methods for carrying out regroupings. The dynamic cloud method distinguishes the nuclei of groups and their fuzzy margins: by this means one rediscovers the results to which superimposing analytic maps led.

The use of factor analysis

One can still calculate the distance separating the points showing primary areas in a space of n dimensions: the method automatically leads to results, but its use is artificial when the characteristics do not vary in the same direction, as we saw in the case of figure 3.4b. Instead of conducting a regrouping of related demarcations by superimposition, one can employ more rigorous statistical approaches: factor analysis or the analysis of correspondence is very appropriate.

Analysis by principal components allows one to distinguish, within a suite of statistical distributions, those which are correlated and can be represented by a common factor. Individual components are then projected on each of the factors and the scores obtained can be mapped. One can thus discriminate the spatial distribution peculiar to each of the independent variables which the analysis produced: a result is then available that is more systematic than that obtained traditionally, and neater, because every correlation between the components successively shown has been eliminated.

Analysis of the correspondences may also be used. In this case it is possible to transfer the various characteristics and individual points to factorial bases, groupings in the form of clouds take shape. The

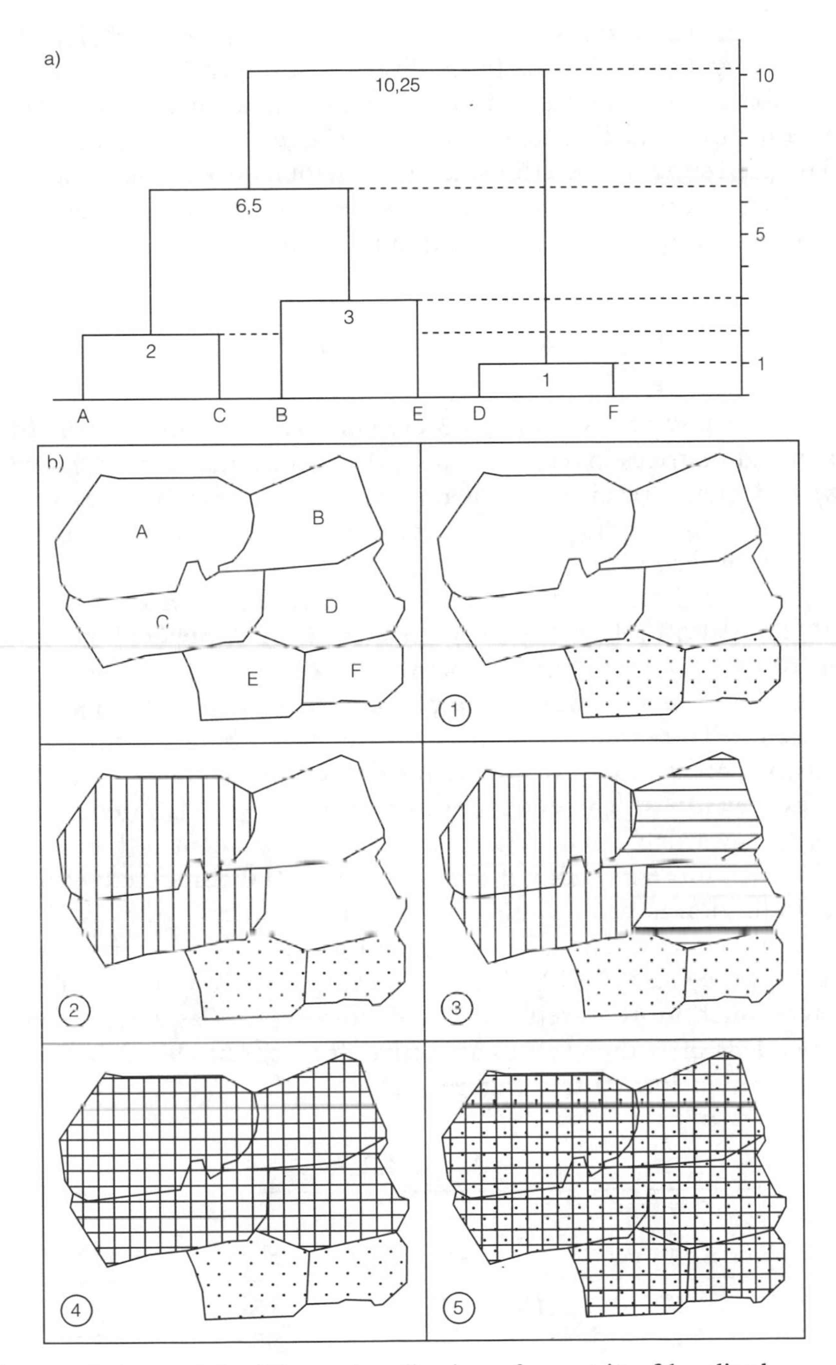

Figures 3.6 a and b The regionalization of a matrix of localized observations

presence in the heart of a cloud of a certain number of characteristics allows its significance to be given. In this way one regionalizes by combining the strictness of the initial statistical analysis and the intuitive nature of distinguishing between classes on a diagram.

The perfecting of powerful statistical methods has made for a better understanding of the logical nature of the concepts of regionalization, and allowed a more effective handling of large data sets.

Defining nodal regions

Statistics of flows between regions are generally presented in the form of squared matrices in which what leaves each spatial unit is shown in rows and what it receives is shown in columns. These matrices are not symmetrical, as the flow returning to a point may be higher than that leaving it or vice versa.

Matrix 3.2 therefore describes the relations between six points, A, B, C, D, E and F. The larger centres attract the strongest flow: as the total of the columns (entries) demonstrates. A receives the highest quantities and is thus one of the points around which the network is articulated. To see how relations are established between the pairs of towns, we must next turn to the rows: the highest value of each indicates towards which centre the town it represents is oriented. A is oriented towards C, but C is a far more modest centre than A, and sends much more to A that it receives from it (40 compared with 25). This indication is without interest therefore. But it is towards A that the dominant relations of C (40) and F (33) go. For its part, E attracts B (35) and D (40). E sends its strongest flows to B (30), but receives much more in return (35). E thus appears as a second pole, independent of A though less attractive. The ensemble of these links of polarization form a tree (figure 3.7).

Matrix 3.2

Exits / Entries	A	B	C	D	E	F	TOTAL
A	–	15	25	20	18	22	100
B	30	–	30	15	35	20	130
C	40	17	–	20	25	13	115
D	25	30	20	–	40	05	120
E	22	30	20	20	–	13	105
F	33	20	18	15	20	–	106
TOTAL	150	112	113	90	138	73	–

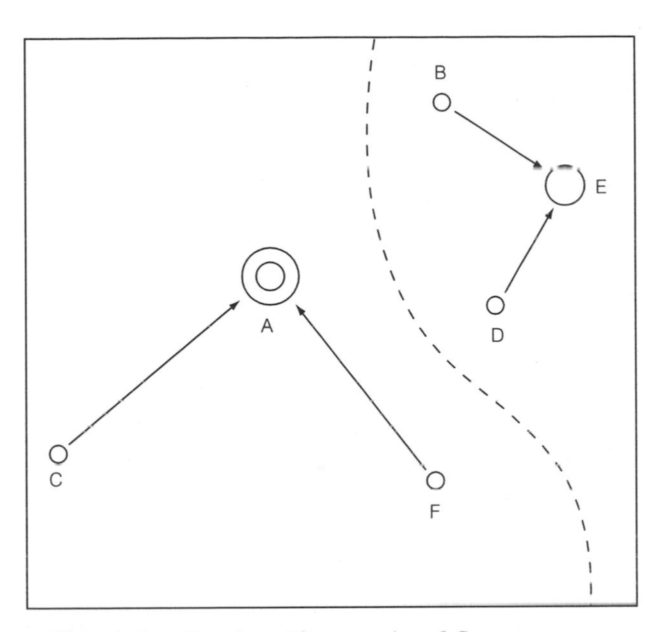

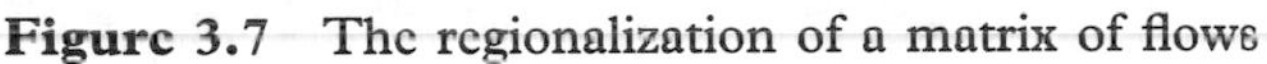
Figure 3.7 The regionalization of a matrix of flows

There are therefore simple procedures for extracting from a matrix of linkages a picture of the facts of polarization, which may then be mapped.

Conclusion

The effort of analysis adopted by geographers anxious to understand regional differentiation of the land is very complex, because it targets various objectives: they wish to describe the diversity of places, explain the structuring of natural milieux and the spatial organization of human activities, understand the experiences that people have of the local or distant environment, and help to choose divisions that are most propitious to economic development, the exercise of democracy or cultural creativity. There are many correlated approaches, but they all have one point in common: they imply that divisions must be made and boundaries established within the area under study.

The operations leading to those divisions were for a long time empirical, but recent consideration has allowed them to be given a more systematic basis. That was essential in order to strengthen overall regional research.

PART II

FACTORS OF THE DIVISION AND REGIONAL ORGANIZATION OF SPACE

4

The Ecological Foundations of Regional Geography

In order to understand how an area of land is organized, how it is broken up into small units, how these are structured and how they combine into much larger regions, we must concentrate on an overall study of the flows of energy, matter, living things, people and information that develop on the surface of the earth. Here a fundamental distinction must be made:

1. at any point, a flow is established downward or upward between the atmosphere, the bedrock, the subsoil, the soil and the plants and animals it supports;
2. exchanges and relations develop laterally between elements situated in different places. The world is thus woven together by both vertical and horizontal connections.

Vertical and horizontal flows interact. In certain cases, the complex whole of what they set in motion circulates within limited compartments: structuring in land units that are independent of one another is impossible. An infinite number of juxtaposed points has been replaced by a restricted number of regions. Each may be studied by itself, since there is interaction only within it. Regional geography relied on this at the beginning of the century: those who practised it felt that they could abstract from their context the essential parts that interested them.

For the most part, horizontal flows embrace ensembles of varying size, so that they are not contained within the same confines. Regional analysis no longer consists of pinpointing autonomous units while overlooking what is taking place outside their boundaries. Rather, the intention is to grasp, within an ensemble subject to many external influences, what makes it react in a distinctive manner; geographers therefore dwell upon the particular architecture of the flows that characterize it. There is no longer any radical distinction between the region and the area that surrounds it: subject to the same forces, it does not however respond to them with the same intensity as neighbouring zones, and is affected by them for longer or shorter periods.

4.1 Natural Interactions on the Earth's Surface

Regional differentiation on the earth is due to the influence of natural phenomena and the action of human societies. The environment is not simply scenery or a support: it imposes constraints on the groups that have settled there, but in return is subjected to pressures, that are often severe, and to major modifications. The disruptions created by industrial societies are such that nowadays they threaten important biological balances.

The physical framework and morphological viewpoint

The plains, mountains, valleys and hills that form the backdrop on which the lives of humankind are inscribed have their origin in the tectonic forces that positioned them and the processes of erosion that attack them. These phenomena have occurred over a time scale that has little in common with the measure of the human lifespan. For individuals as for societies, the terrain, the nature of rock outcrops and the general relief are stable data. Changes in the topography, which are virtually incessant, are normally imperceptible. Some zones shift more: they are the theatres of violent phenomena, earthquakes, volcanic eruptions, devastating mudslides, etc. These recurrent natural catastrophes affect clearly defined areas, where their effects are cumulative; even there, on a scale of generations, stability generally prevails. The potential danger does not prevent people from living in large numbers on the slopes of Etna, despite volcanic activity, or in Japan despite earthquakes.

Morphologists make distinctions between a few major types of unit according to their relief and geological structure: young mountains, ancient massifs, sedimentary basins, etc. It is a classification that suits the problems they research: it takes account of the composition of rocks (useful in assessing erosion) and the strength of the relief (linked with tectonic forces). For the geologist in search of metalliferous ores, altitude and volume count for little; what matters to them is to find old basement rocks, whether it is at sea level or at an altitude of 4,000 metres. Those prospecting for oil deposits are interested in deep sedimentary basins, whether they have developed inland or on the continental shelf. There is no universally valid classification of the topographic frameworks in which human activity takes place. Their relevance depends on the way in which they are used. For the crowds of tourists, the quality of snowfall in the mountains or the abundance of fine sand on a sunny shore are the things that define the contours of a valued milieu.

When one conducts regional analysis, what matters above all is the general framework offered to the activity of human societies. Low, flat regions, where working the land and getting from place to place are both easier seem *a priori* preferable to the mountains. But that is far from being confirmed everywhere. In hot zones especially, the higher regions are often more favourable: they are more humid in arid land, healthier in a wet environment, with terracing that offers a range of possibilities and topography that frequently creates better conditions of security.

To describe the framework in which the life of humankind is inscribed, the morphologist's viewpoint is too narrow. It allows an appreciation of the facilities offered to the life of interrelationships: the arrangements of the lines of relief, the altitude and the continuity of mountainous settings are constraints that, despite technical progress, have an influence on movement and trade. But for other aspects of human activity, one must take account of climates and all the processes that connect the living world with the mineral world and the cosmos.

The ecological viewpoint

Nature is not an inert environment: many interactions take place within it. First, they bring into play the atmosphere and the lithosphere. Rocks attacked by wind, rain, heat or frost develop fissures, soak up water and disintegrate. The soil is then ready to receive seeds and allows root penetration. Plant and animal life develops. All these living creatures die and the organic matter formed by the humus slowly becomes mineralized. The thin layer of soil thus created becomes the intermediate milieu that is essential to pass from the mineral to the living world.

Earth is a living planet. The biosphere of which we are part developed in contact with the geosphere and the atmosphere. The conditions that permit life to flourish vary greatly from place to place. The natural regions that interest us are the botanists' biotopes.

The sun's rays supply the necessary energy for the transformations that occur on the earth's surface. Without them there would be no photosynthesis, no production of organic matter, no plants and no animals. The elements composing the environment are connected by flows of matter and energy and constitute *ecosystems*. This is the name given to the complex biological systems formed by various organisms living together in a given area and by elements of the environment which impact on their existence. Their extent varies; they give an area its most fundamental forms of structuring.

Green plants perform the synthesis of hydrocarbons: to this end, the solar energy that their leaves transform into chemical energy thanks to chlorophyll allows them to make use of the carbon dioxide in the air and what is derived from their sap. From the subsoil their roots draw the water and the various mineral ions necessary for the building of living tissue. The insolation received supplies both the energy necessary for chemical synthesis and that consumed by the rising of the sap by evapotranspiration. Plant products provide food for a whole range of herbivorous and carnivorous animals (food chain) and micro-organisms.

The organic molecules are partly burnt by the living creature itself (respiration), which draws from them the energy needed for it to function. From one level to the next in the food chain, the living matter produced thus dwindles. The whole constitutes an ecological pyramid. In order to produce this sequence, certain conditions must combine: there must be insolation, water and a soil which, by its structure as well as its richness in useful ions (absorptive complex), allows the efficient functioning of exchanges at root level. If one of these elements is insufficient or missing (limiting factor), the process is halted. The productivity of plant associations and the entire animal pyramid they feed is sharply lowered.

In aquatic environments, the chlorophyllian synthesis for which (often microscopic) algae are responsible is similarly the basis of the production of living matter. Oxygen (dissolved in limited quantities) and the concentration in ions, necessary for the synthesis of the living creature, constitute the limiting factors. As the chlorophyllian assimilation does not rely on evaporation, it is less sensitive to temperatures than on land: productivity is even higher in cold waters, for they contain more dissolved gases. However, it is never equal to that observed on land areas because the concentration of fertilizing matter is too weak.

4.2 Ecological Mechanisms and Space

The kinds of circulation that allow plant life to grow on land are first expressed in the vertical dimension (figure 4.1). The sun's rays and the infiltrating moisture from precipitation descend; directly reflected or infra-red radiation, evaporation, the rising sap and evapotranspiration are directed upward. The compartments in which the transformation

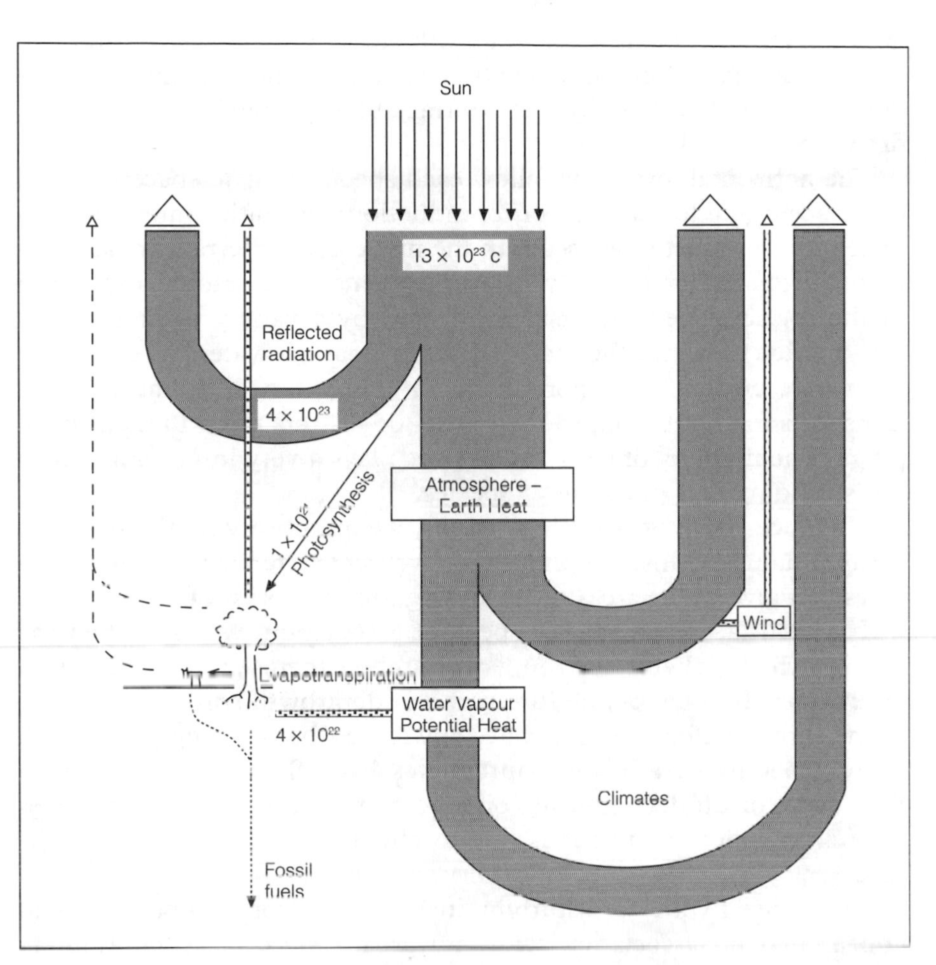

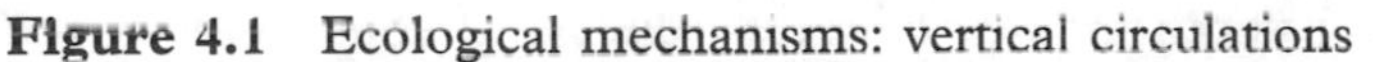

Figure 4.1 Ecological mechanisms: vertical circulations

The terrestrial ecosystem functions due to solar energy, partly reflected, partly mobilized by evaporation, evapotranspiration and photosynthesis and partly dissipated in mechanical form – which introduces into the climatic mechanisms a horizontal component (energy is expressed in calories).

Source: P. Claval, *Eléments de géographie humaine*, Paris: Litec, 1975, p. 84.

of energy and matter takes place on the earth's surface are confined. The unit is formed by the space occupied by a plant and its roots. It captures the sun's rays with its foliage, and from the vertical cylinder where its roots spread it draws the water and mineral elements it needs. For the most part, these basic cylinders overlap one another to such an extent that all the soil is made use of in both surface and depth. Interweavings are especially numerous in plant formations comprising several superimposed strata, trees, shrubs and grasses. The plants that compose them do not all cover the same expanse and their roots extend to varying lengths and depths. In arid zones the situation is different: each plant must drain a large volume of soil in order to obtain enough humidity: vegetation is sparse, the sun mostly beats down from a cloudless sky, yet contrary to appearances, living things draw upon the whole area.

The horizontal forces that allow each species to gain space for itself differ according to the form of reproduction: vegetative multiplication or sexual reproduction by seed. In the first case, subterranean development of the rhizome or aerial layering cause the gradual expansion of the species. Seed dispersal is effected by a variety of methods: it may be due to wind, the fact that seeds float on watercourses or the sea where currents transport them, or the action of animals, which transports them attached to their coats or scatters them in their droppings. The mobility of fauna of all sizes plays a very important part in the spreading and mingling of species.

When we speak of plant formations, we are observing the presence of continuous expanses of plants with well-characterized dominant features that give whole areas their physiognomy: a forest of oaks, steppes with artemisia, etc. When we speak of association, we mean that the species which cohabit are connected among themselves by necessary interaction. In a temperate forest, the undergrowth sprouts very early in the spring, before the trees have any leaves; lianas or epiphytes and, more generally parasites or saprophytes live off the plants to which they cling, or off decomposing organic matter. But in many instances the juxtaposition of species is due to chance, the existence of a suitable ecological niche. Deadly rivalry may arise.

Left to itself with no disturbing intervention from people, it would appear that an association tends towards a state of equilibrium in keeping with local conditions of climate, relief and soils. This is the *climax*. Nevertheless, this concept has more theoretical than practical interest, given that human influence everywhere is both ancient and deep even where nature appears to be in a wild state.

Micro relief introduces local modifications that may be evident in the floral composition of a formation: water flowing rapidly away down slopes carries off part of the precipitation. When infiltration occurs, it does so only where the slopes diminish and where the water flow slows down. Water is rarer on the peaks and upper slopes, and tends to accumulate and stagnate lower down. Therefore sequences of varied types of vegetation cover the slopes. Relief creates other kinds of spatial mosaics. The amount of solar energy received by a plot of land changes with its exposure. It is stronger on south-facing slopes (in our hemisphere), as the rays arrive at a sharper angle.

The majority of land animals are mobile, and introduce a horizontal component to the interactions at the basis of the ecosystems. In so far as their movements are random, their wanderings do not appreciably alter the balance of the flows in each of the basic cells that make up the milieu: the animal eats randomly. Its droppings also fall here and there. The resultant balance of imports and exports of matter and energy is virtually nil. If one insists on the minimal dimension that certain natural ecosystems must have in order not to experience disequilibrium in their pyramids of living creatures, it is because some animals need vast areas of space if they are to develop normally. Interactions with a horizontal component are important between fauna and flora. In theory, an equilibrium is established between the two, but it is a fragile one, especially at the level of the larger animals. Herds of herbivores need to roam across immense spaces, depending on the season, to find the grazing they need. The large carnivores jealously mark out their territory to find the prey that constitutes their menu. If they disappear, overgrazing becomes a threat to vegetation, and at the same time to herbivorous species.

The oblique and horizontal components of ecological pyramids are by no means negligible, but it is the vertical dimension that is essential. This is what tends to fragment the earth's surface into minuscule units; this is what causes the slightest nuances in the nature of soils and subsoils, in aspects and slopes, to be revealed in forms of vegetation.

4.3 The Distribution of Ecosystems

Thresholds and zones

Along with the terrain that they carpet, the large vegetation zones constitute the foremost regions in the earth's mosaic. It is at their level that the flows of matter and energy give birth to specific combinations.

The spread of plant life depends essentially on two factors: sun and water. The intensity of the sun's rays controls chlorophyllian assimilation, and temperatures also depend on it. Below 10° C, plant functions slow down. Their water supply is basically connected with local rainfall. The water that infiltrates the soil dissolves the elements necessary for organic syntheses. There must be a certain temperature for evapotranspiration to take place and the sap to rise in the plant as far as the leaves where organic syntheses occur.

The values of insolation, temperature and rainfall are never exactly the same in two neighbouring places. Their variation normally works in a continuous manner. Differences are generally negligible over distances except where the topography introduces marked contrasts in the amounts of insolation and the depth and stability of the water-tables. Ecological pyramids, therefore, are generally identically reproduced from place to place as long as insolation, rainfall and relief do not change too much. This allows us to apply the term ecosystem to extensive ensembles, whereas the majority of exchanges are effected on the basis of very restricted cells.

From the equator to the pole, insolation gradually decreases, thermal range is reduced by the proximity of large masses of water (oceans, seas or large lakes), which diminish its fluctuations. Rainfall dwindles the farther away from evaporation zones when no obstacle occurs to thermal rising, and in zones where masses of air are sinking. The earth's surface ought to present itself as a mosaic of plant formations composed of very small units, all in transition, where it is not possible to recognize large ensembles. What it displays, on the contrary, is a composition in blocs with relatively clear outlines, which extend sometimes over millions of square kilometres, like the Siberian or Canadian taigas, the American prairies, the Amazonian forests or African savannas. In places where the relief is fairly subdued and geological structure homogeneous enough not to interfere with the influence of the latitude on insolation and the action of atmospheric movements, what is most striking is the predominance of vast uniform stretches of landscape. The variations in the factors that influence life do not affect the functioning of living ensembles as long as they remain contained within certain limits. The forms that depend on them remain similar as long as certain thresholds are not crossed.

Continuous variations in the intensity of factors give rise to effects that remain constant for a long time; everything changes once a critical value is reached. Tendencies to homogeneity and tendencies to differentiation thus merge: ecosystems keep the same structure as long as the factors that influence them stay within a certain range; they change their physiognomy as soon as the threshold is crossed. Where relief is uneven, slopes, altitude and orientation come into play and different limits may be crossed over short distances. Thus one sees the appearance of vertical layering in mountain areas, contrasts between south-facing and north-facing aspects, or leeward and windward slopes.

Places where one is far from critical values for all the limiting factors, insolation, rainfall, temperature or type of soil, topographical effects are not sufficient to cause the threshold to be crossed: formations scarcely reflect the local relief features. The rainforest is continuous in the heart of the Amazonian basin or the loop of the Congo. But when their limits are approached it takes a very minor change for the critical value to be reached: topography then becomes an essential differentiating factor (figure 4.2). On the north and south fringes of the forests where evergreens are dominant, the forest galleries covering the damp valley floors extend the main forest cover while the interfluves and slopes are already given over to savanna.

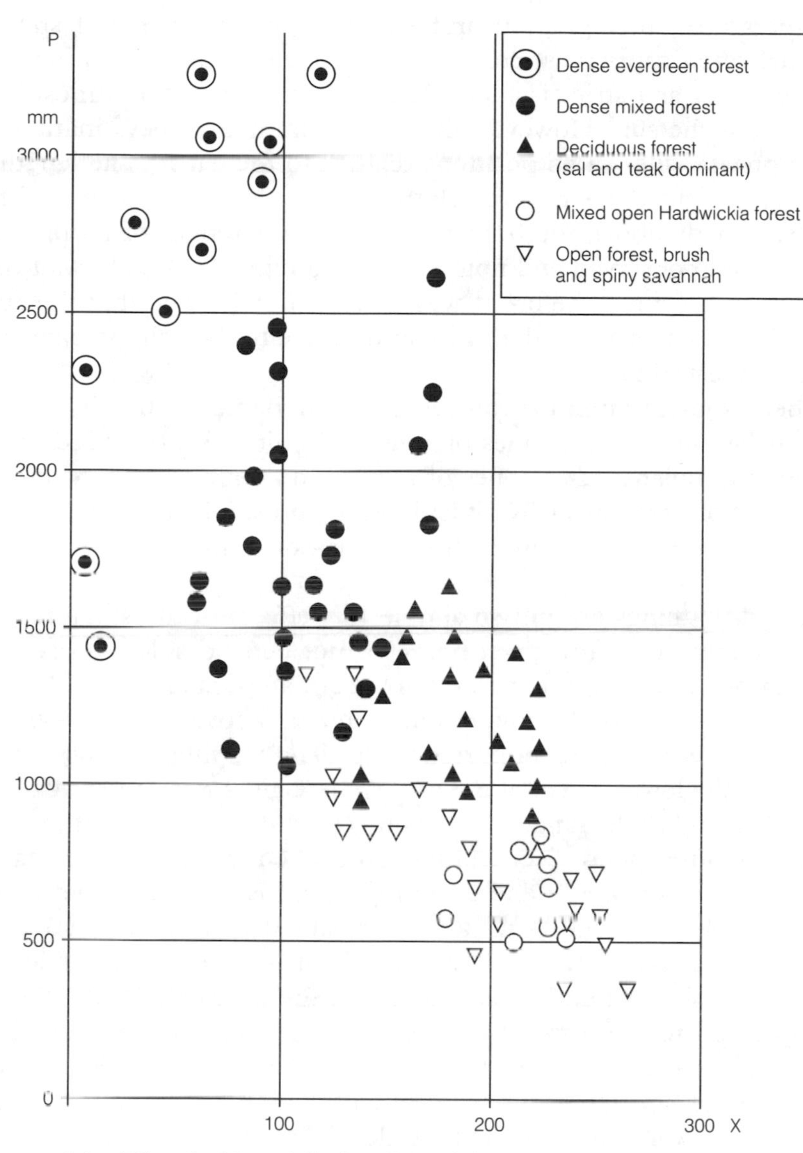

Figure 4.2 Threshold and limit effects in vegetal formations
Correlation between precipitation (vertical axis), the xerothermic index (horizontal axis) and the principal vegetation types of India and Ceylon after Legris. Clear thresholds exist between the first four types. The fifth partially overlaps with the other domains.
Source: P. Birot, *Formations végétales du Globe*, Paris: SEDES, 1965, p. 413.

The major biotypes: climates, natural vegetation and soils

The essential factors differentiating ecosystems must therefore be sought in climate and relief. Climates are familiar realities but difficult to define.

Temperatures, precipitation and winds appear to be manifestations that are often capricious, constantly changing, hazardous, impossible to predict in any strict fashion. The great motor of the atmospheric machine is the sun. However, the latter immutably obeys mathematically precise rules in its positions relative to the earth. The length of days and nights, the rhythm of the seasons and the height of this heavenly body above the horizon control the strength of its rays. All this has been observed and noted since antiquity. We had to wait until the fifteenth century before it was admitted in Europe that the rotation of the earth around the sun and its angle to the ecliptic explained all the observed facts.

There is justification for considering as fundamental the divisions of the earth's surface into zones of similar illumination, bounded by the tropics and polar circles: zones of arctic and antarctic cold, where living conditions are especially difficult, even impossible, temperate zones, marked by the alternation of thermal seasons, and hot intertropical zones.

In reality, things are not so simple: the effects of the sun's rays on the temperature at any given point do not depend solely on latitude and season. The hygrometric level of the air, amount and type of cloud, nature and colour of the soil, orientation and, above all, disposition of lands and seas play an important role. The varying heating experienced by the lower layers of the atmosphere gives rise to unequal pressures and a whole range of turbulence of differing amplitude that mixes the atmosphere. Vertical movements contribute to altering the pressure of water vapour and encouraging or preventing condensation. In zones of subsidence, the air heats and becomes desiccating. In areas of low pressure, it rises and cools down as the pressure is released, causing the formation of fine droplets in the heart of clouds, then of precipitation in the form of rain or snow. The heavy, cold air of the polar regions comes in great waves to cool the temperate zone regions and even the borders of the tropical world. The hot air of the tropics also edges towards the middle latitudes. When these major air masses make contact, violent thermals are produced, marked by winds and rainfall. The unpredictability of the fronts' movements largely explains the caprices of the climate.

Nevertheless, large ensembles take shape, separated by thresholds that reveal themselves in changes in the ecosystems. Limiting factors are not the same everywhere.

(1) In the polar and subpolar latitudes, temperature rules everything. Precipitation is slight as evaporation is negligible over the water bodies (frozen for months at a time) and the quantity of vapour in the

atmosphere is always low. Precipitation is in the form of snow, which does not melt and piles up on the land in enormous ice caps that ultimately overflow into the neighbouring seas. Only a few living things are known there, and solely on the edges. Where the land is low-lying and free of ice and the ice floes recede for a few weeks or more in the summer, abundant phytoplankton develop under the low-angled but contiguous light of the arctic summer in the cold and well-oxygenated waters. It feeds a whole range of marine animals, the higher echelons of which, especially the seal, have provided the basis for the way of life of Eskimo societies. On land, the intensity of the cold is less of a determinant than its duration. The summer thaw is not sufficient to penetrate very deeply. The permafrost forms a hard layer that is impermeable and impenetrable by roots. Decomposition of organic matter and the formation of humus take place with difficulty or not at all. There are purely mineral soils that suffice for a few patches of lichens, and also peaty organic soils. The vegetation of the tundra forms intermittent expanses, marshes, grassy areas and very low-lying scrub. In spite of everything, when the summer arrives, it is enough for a range of large mammals to appear, or small rodents, which serve as prey for foxes, wolves or bears. People have managed to live by hunting in these regions, and by breeding deer in Eurasia.

(2) In the middle latitudes, a wide variety of situations can be seen. The duration and intensity of the cold season interact with the variable amount of rainfall to produce a mosaic of habitats.

From the point of view of temperature, it is the duration of the vegetative season (average monthly temperature over 10° C) that is crucial. Where it normally does not exceed four months, predominantly coniferous northern forests occupy the area. Trappers and loggers have traversed and to some extent plundered these regions, but farmers have been discouraged: just as much as the harshness of the climate, the podzolic soils and mediocre drainage, as revealed by the abundance of peatbogs and swamps, have prevented any normal agricultural development.

As one goes southward, the length of the warm season increases and the severity of winter diminishes. Conditions become favourable to formations of deciduous trees with a winter rest from vegetal growth: this is the temperate forest, dominated by oaks, which becomes the principal vegetation type as long as rainfall is adequate. Where precipitation declines on average below 600 or 500 mm, in the land masses of Eurasia or the inland plains of the United States, Canada and Argentina, the forest thins out to give place to moist prairie, which deteriorates, in parallel with dwindling rainfall, to dry prairie (the Russian steppes) and true desert in enclosed basins.

(3) Around the 35th parallel, winter warmth is sufficient, during the month with least sunshine, that temperatures above 10° C on the western façades of mainlands can still be reached. On eastern façades, the same transition is seen further south, around the 30° parallel. The thermal seasons therefore lose their significance and it is the level of precipitation which now constitutes the limiting factor.

In these latitudes the general conditions of atmospheric circulation cause a striking asymmetry between the east and west façades of mainlands. On the western façades (Mediterranean region, California), a narrow zone with slightly moist winters and dry summers, marks the transition with the arid tropical world dominated by circulation from the west. As one descends in latitude, the desert rapidly takes over, from the coastal regions as far as the continental interiors.

On eastern façades, (China, United States), summer rains of the monsoon type ensure the penetration inland of sufficient moisture. Then one traverses imperceptible transitions from the temperate world with cold winters, with its procession of conifers and deciduous trees, to tropical forest. The great richness of the flora in these regions contrasts with the monotony of European woodlands that essentially lack much variety.

The subtropical and intertropical regimes are generally much more clear-cut than those of the temperate zone. The often torrential rainy season is followed by a particularly hot and dry season that lasts a relatively long time. In those regions that are well watered all year round, near the equator (Amazonian basin, Guinean zone and the Congo Basin in Africa, Malaysia and western Indonesia), the luxuriance of the plant world is extraordinary. Judging by the number of plant species to be found there, the height of the trees, the complex terracing of layers of vegetation, the teeming animal life, these great forests have nothing in common with European forests. Where there is a dry season, forests of deciduous trees replace the evergreen jungle, but they are more fragile and also more under attack. The savanna with its tall grasses then replaces them in part or completely: this is the domain of the large herbivores and their predators, chiefly in Africa and still to some extent in India.

These regions which appear so favourable to life are actually difficult for humankind. Germs proliferate and the state of health of populations and their livestock is precarious. When the ground is cleared for cultivation, soils are shown to be fragile, leached, not very fertile – in short, mediocre. Agriculture, however, appeared there very early: the forest environment, in fact, lends itself well to root propagation. Tropical horticulture is as old in New Guinea, for instance, as dry-farming in the Middle East.

Vegetation landscapes shaped by man and by natural conditions

The variations in living landscapes have for a long time been disturbed by the actions of humankind. The Australian aborigines were few in number and knew nothing of agriculture; consequently, their effect on the vegetation cover was reduced to a minimum. Nevertheless, it was far from negligible. To meet their hunting needs, indigenous tribes were in the habit of periodically burning the areas in which they had settled. Two centuries ago, when the Europeans arrived, the formations dominating the continent were of species that could withstand flames. The dense forests spared by the fires were rare outside the wetter zones of the south-west, south-east and the eastern coastal strip. Everywhere else, the country looked like a park dotted with isolated trees: the result of repeated fires.

In the majority of cases, even when they are not cultivated, the landscapes that we see before us owe still more to the actions of humans. Afforestation, pasturelands, meadows and land under cultivation, like purely natural formations, are reliant on climatic and soil conditions. These formations are also characterized by the succession of plant and animal species that the cultivators have not been able to eliminate when destroying the original formations, or that they have introduced with the seeds they have planted.

For living landscapes controlled by man, threshold effects exist in exactly the same way as when nature is left to itself. The major boundaries between plant formations generally retain their significance. In the Ivory Coast, the cocoa and coffee plantations that have ensured a quarter century of prosperity for the country were installed on cleared forests. The savanna did not allow a similar development in cash crops, and the gap in the levels of life and amenities between the north and south of the country rapidly deepened.

In cold temperate countries, agricultural exploitation goes no further than the mixed forest that lies between the taiga and the broadleafed forests. This has to do with the climate: the risk of frost in spring or autumn worsens rapidly towards the north, and the length of the growing season becomes too short for the majority of cultivated plants, especially cereals. The mediocrity and acidity of the podzols also discourages farmers.

The thresholds defining zones where crops are possible do not always follow the line of those defining areas of natural vegetation. This is explained by the transformations which farmers can exercise, and the manner in which, by selection and hybridization, they modify the species on which they concentrate their efforts. With the use of fertilizers, the chemical composition of soils has in many instances ceased to be a limiting factor; calcium enrichment and doses of nitrate, potassium and phosphorous bestow on poor lands qualities that are close to those of regions reputed to be fertile. All that is required is to offer soils with a favourable texture! Brittany or the province of Dry Champagne no longer deserve their reputation of poorly endowed areas, and support intensive crops.

Does this mean that the tyranny of climatic constraints is strengthened? In one sense, yes: whichever area is under consideration, there are always crops that average or extreme temperatures and insufficient or ill-distributed rainfall rule out, but the limits are by no means fixed. Species that are the outcome of selection or hybridization often stand up better to difficult conditions than those they are replacing. Since its introduction into Europe, maize had been a crop for regions with hot, moist summers: Aquitaine, the Po valley, the valleys of the Saône and Rhine, the basins of central and Balkan Europe. The hybrids from America in the 1950s caused this plant to advance 1,000 km northwards. Today it flourishes in all the French plains, and extends as far as the Hamburg region in Germany. In eastern France, it could not be grown except on the plains. Today it has become one of the key features in the systematic rotation of crops in part of the Jura mountain chain, where it has drastically changed livestock rearing. To an altitude of 600 or 650 metres, maize as animal fodder has replaced what was required from other cereals and root crops, and has allowed rapid intensification. At a higher altitude, the old pastoral systems that make wider use of grass have not been altered. The foothill plateau was regarded as being poorly endowed when compared with the mountain and its rich pastures. The intensification following the introduction of maize upset the balance.

In many instances, the inadequacy or poor distribution of rainfall formerly hampered the extension of crops. There, too, the thresholds have been displaced. People have managed to adapt wheat to dry regions by selecting varieties the growth of which coincides more precisely with the period when moisture is sufficient. Their stalks are shorter, which allows them to bear more grain without the necessity for greater evapotranspiration. Elsewhere, irrigation is called upon to provide the essential complements for plants.

The geography of cultivated plant associations is always characterized by the existence of wide areas where conditions do not change enough to give rise to variations, and by thresholds where new combinations are suddenly imposed, but the limits have ceased to be settled once and for all: genetic engineering and the techniques of fertilization and irrigation allow a substantial freedom from the constraints of nature, and the progress that constantly asserts itself is revealed by a continual remodelling of the landscapes under cultivation. Is everything possible in this domain? No: enriching, irrigating or using selected varieties, which are more productive but more fragile, carry a price. Vegetables may be produced under plastic tunnels or in heated greenhouses even in the most extreme environments, but the price to be paid is so high that they become a curiosity that is inapplicable to a wider geographical scale.

4.4 Ecological Mechanisms and Regional Organization

The way humans fit into ecological pyramids: the biological dimension

Like all living things, men and women are included in ecological pyramids. The energy necessary for their organism to function is supplied by the plants and animals they consume: they are integrated into trophic chains. According as to whether their diet is vegetarian or meat-eating, men and women are placed, in these sequences of transformation, on the level of herbivores or carnivores: they make use of one or several strata of the living world. The share they receive varies, for they are always competing with species that are in rivalry with them for their fundamental food requirements. They widen their ecological base by replacing natural growths with crops the harvests of which they can more easily assimilate and by eliminating the animals that plunder them.

The place occupied by humans in ecological pyramids depends on the productivity of the pyramids and the way in which people govern their use. The proliferation of parasites, weeds or harmful plants often dramatically reduces the amount they are able to take out. The position of human beings also depends on their state of health: consumable products are used by others if their population cannot multiply, or is dwindling. The first condition for remaining fit is to have an adequate diet, and one which contributes in good proportion to the carbohydrates, lipids, proteins, vitamins and mineral salts that are indispensable to the equilibrium of the organism. When some of the elements are lacking in the environment, or people are unable to procure them, they find themselves exposed to deficiency diseases. Their incidence varies: regions where nourishment relies too exclusively on cereal-growing have populations which suffer sometimes from illnesses linked with lack of proteins – especially dramatic among children who are then afflicted with marasmus – and with lack of vitamins, beri-beri among eaters of husked and polished rice, or pellagra when maize is the basis of all foods.

The human organism can also be weakened by parasites, or attacked by bacteria and viruses. Their spread depends on the density of population and the intensity of human relations. It is also the work of the natural world: intestinal infections are all the more frequent when the water one drinks is contaminated by excrement. The incidence of respiratory ailments is all the greater when the sun, that great bactericide, visits dwellings less often. Certain milieux lend themselves to the perpetuation and propagation of harmful germs: in this regard, geographers speak of pathogenic complexes. The geography of human settlements reflects these conditions: densities are often lower and populations less vigorous where health conditions are not assured and where ailments are especially frequent. Marshy regions were long regarded as repellent, but methods of destroying mosquitoes and the use of nivaquine gave normal living conditions to the populations that lived there during the 1950s and 1960s. For fifteen years or so, their position has been deteriorating: mosquitoes have developed a resistance to certain pesticides, and other pesticides, such as DDT, have been banned because of their harmful effects on the environment as a whole; the malarial strains are no longer sensitive to the pharmacopoeia that caused such a sensation forty years ago.

Medicine's progress is so spectacular that one often tends to forget the biological dimension of human inclusion in the environment, but many hot and moist regions remain very unhealthy.

Needs in raw materials and energy

The relationship of groups with their environments also needs to be analysed from the point of view of exchanges of raw materials and energy. People draw all their food and part of their drink from plants and animals. For millennia they extracted the fibres they needed to clothe themselves, and the wood which gave them warmth, served to make their tools and contributed (and still does, often predominantly) to the construction of their shelters, huts, cabins or houses. Unlike animals, people are not satisfied to take from their surroundings the amount of calories matching their metabolism. Long ago they learned to isolate themselves from the environment by clothing themselves and building dwellings for themselves and sheepfolds, cattlesheds and stables for the animals they bred, stores for their own provisions and those necessary for their flocks and herds. To make their tools effective, they used metals derived from ores they found in the subsoil. The energy that the human body is capable of providing could not meet all those needs: from the Neolithic Age, people of the ancient world, from Europe to China, knew how to make use of the strength of their domestic animals to carry or drag loads and to turn wheels. It was then that a gap in technical efficiency opened up between the peoples of sub-Saharan Africa and those of pre-Columban America.

Human groups are included in ecological pyramids, where they draw on far more varied elements than do the animals. The list of what they borrow from the living world was not established once and for all: it grew longer with the progress of civilization until the moment when the use of fossil fuels was mastered. Human relations with the environment were altered by this: until then people had known how to employ only energies derived from the sun which were generally scattered like the rays that gave them birth: the chemical energy of foodstuffs, the mechanical energy of winds and running water. This limited the possibility of people concentrating together.

Today, as in the past, food comes from the animal or vegetable kingdoms, but energy is obtained from concentrated sources. To a small extent it is provided by the hydroelectric plants, and in the main by coal, oil and natural gas. Nuclear reactors are increasingly taking over. The functional world is drawn from the cultivated plants, domestic animals and forests; it is increasingly built on what is extracted from mines and, thanks to chemistry, from oil and gas.

Human beings are always torn between the demands of the life of interrelationships and the need to take from the environment their means of subsistence, the raw materials and the energy they use. The way in which they split up into independent units and structure the land is the result.

The problem of food: human groups and their ecological support

People are capable of transporting what they need over long distances. This considerably modifies their insertion into the ecological pyramid. Animals move in order to find and eat their food and do not trouble to take it back to where they shelter unless they have young to feed or, in certain instances, to pass the winter. Humans establish a systematic distinction between where they live and where they take the products they consume or commercialize. The moment they exploit ecosystems, the latter cease to comprise vertical circulations alone: a horizontal component is introduced (figure 4.3).

Humans take what they consume from fairly extensive surfaces, transport it to the place where they have built their homes or to the market that opens up more distant destinations. The ecological cycle ceases to be enclosed. This creates a fundamental problem – how to effect replacement. The pyramids of living things normally recycle *in situ* mineral elements incorporated in the organic matter: when plants

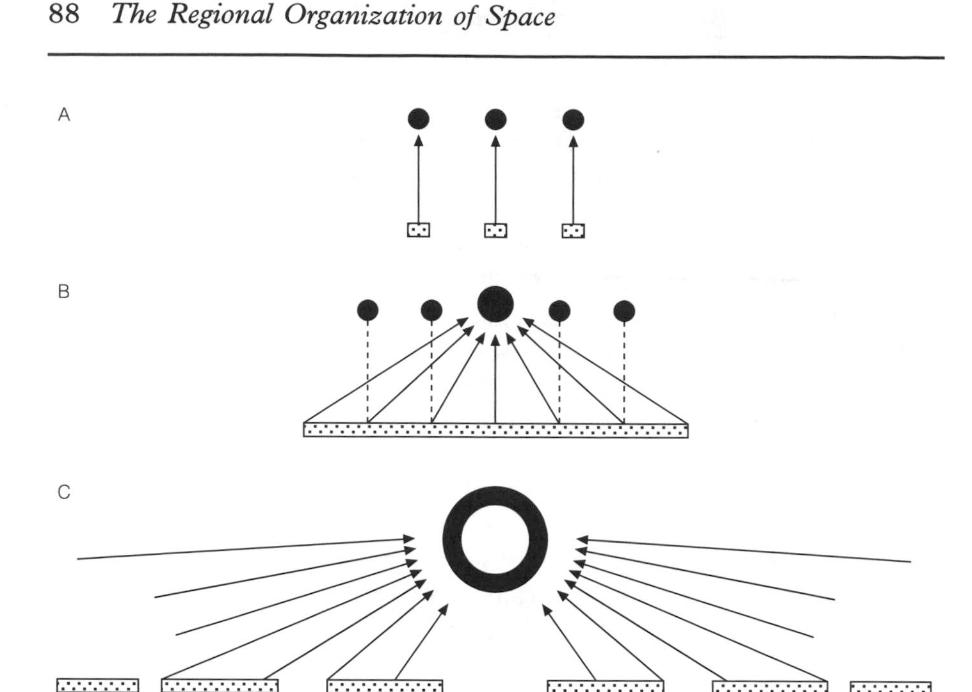

Figure 4.3 The 'horizontal' component in humanized ecosystems

A In primitive societies, small groups depend entirely on local resources: the relationships with the environment are 'vertical'.

B In intermediate societies, peasants continue to depend on self-sufficiency, but towns are nourished essentially by the nearby regions: their ecological support is 'oblique'.

C In the modern industrial world, the ecological support extends to the entire world: the ecological relationships have become 'horizontal'.

Source: P. Claval, *Eléments de géographie humaine*, Paris: Litec, 1975, p. 106.

and animals die, decomposing agents take on the mineralization of their tissues, and the calcium, potassium, phosphorous or metals, traces of which are necessary for organic synthesis, can be made again by the use of roots.

When farmers break this endless round, they have to find means of compensating for what they are extracting from the soil. Clearing destroys natural vegetation and the recycling of elements that characterize it. It is in the permanent evergreen forests of the humid tropics that the impact of the transition from natural formations to fields can best be measured: the profuse vegetation often gives the impression of rich soil. Once the cover has been destroyed to grow crops, it takes only two or three harvests to exhaust the content of the burnt forest's ashes. Yields collapse: the vegetation had lived on a closed circuit, so to speak, taking scarcely anything from the subsoil. Soil and climate conditions are not everywhere as rigorous as in the equatorial domain, but the consequences of a break in the replenishment chain are universal: the fertility of land diminishes and the ecological pyramids lose part of their productivity. No society can occupy territory on a lasting basis unless the problem of putting back what has been taken out is solved.

A return to natural vegetation allows reserves to be rebuilt: for a long time farmers relied on leaving land fallow to avoid exhausting it, but that reduced the proportion of the environment that was effectively cultivated in any given year and put the limit on population densities. Each time the rest period of cultivated areas was successfully reduced without lessening their fertility, the number of people living from the ecological support could therefore grow. As long as the places where products are consumed are not far from the fields, it is easy for the latter to be spread with the resultant animal (and human) waste. The loop of circulation of matter is almost reconstituted. But it is by juggling with crop rotation and its connection with livestock breeding that the most dramatic results are achieved. The use of chemical fertilizers is a liberator from these traditional constraints, but it is possible only if transport costs are low enough.

The world population map still reflects the distribution of zones where people were the earliest to take advantage of the almost continuous, or continuous, use of the same soil: people were able to congregate there before the lowering of transport costs allowed a real separation of production and consumption points. The Near East, the Mediterranean world and Europe owe their early expansion to cereal-growing that was soon associated with livestock. By leaving lands fallow and using animal manure, restoration was ensured. Sheep that had been grazing on stubble fields or scrubland were folded in the evening on the ground intended for sowing a cereal crop. Cattle manure was collected from the cowsheds and spread on the land before ploughing.

The world of rice-growing is very different. Maintained fertility is ensured by irrigation. When water is drawn from silt-laden rivers it may bring fertilizing elements, but it is more to the action of algae which fix a certain quantity of nitrogen, to the frequent tillings and the concern to carry out all possible natural restitutions, that the system owes it success. Allowing continuous crops, in southern and eastern Asia, rice fed the densest populations in the Old World. In the New World, maize and potatoes allowed numbers to multiply from Mexico to Peru. Elsewhere, really intensive systems flourished only in confined areas – certain highlands or deltas in Africa, or the mountains of the interior of New Guinea, for example.

Where humans accumulate and consume, their waste products are so plentiful that it is sometimes difficult for the surroundings to recycle them. This has already been noted in traditional rural civilizations: around houses the proliferation of nitrophile plants was remarkable.

Imperfect recycling was also marked by the multiplication of species that attack provisions and all the parasites and micro-organisms that are a threat to the health of animals and humans. Surface water and that of the water-table were polluted by excrement and refuse.

Groups with local ecological support: vertical or oblique relationships

In traditional societies people were not satisfied to make use of spontaneous natural associations. In any case what they would have gathered from them would have been ridiculously meagre: the greater part of the vegetable or animal matter was not suitable for human nourishment. So humans set about natural formations with fire and the plough and substituted pyramids which were often less productive, but of which they could consume or use a far more sizeable amount. By clearing the ground and replacing natural woods or pasturelands with orchards, sowing grasses or planting tubers, humans succeeded in homogenizing their surroundings. In place of the anarchic mosaic that arises from competition between trees, the fall of those uprooted by wind or dead of old age, with an irregular topography, temporary pools or lasting ponds, they managed, by dint of repeated ploughings, removal of stones and cartage, to render pieces of land homogeneous. It was then easy for them to create uniform combinations that they could look after and harvest without too much fatigue.

Transporting the earth's products in one direction and compensating for them in the other consumes a great deal of energy. As long as the latter came from human beings and their domestic animals, the radius of these movements was thus limited. Everywhere that livestock breeding has an important place, the fodder plants necessary for cattle and the resultant manure represent such a large volume that the distances over which they have to be carted need to be kept to a minimum. For animals raised for food, milk or wool, the simplest way is to keep them where the grass grows. That leads to large-scale movements according to the season: nomadism and transhumance provide extreme examples of this. In the mountains, flocks follow the vegetation, going up to the *mayens* in spring and the *alpages* in summer, coming down again when it gets cold and spending autumn on the valley benches where there is still some provision of hay; in the winter, when the growth of grass is interrupted and there is no more fodder to be had 'on the hoof', animals must be fed in indoor stables: their feed cannot then come from beyond 2 or 3 kilometres on the flat, much less over uneven ground. The horizontal component incorporated into ecosystems is by no means negligible, but the whole of the elements that are included in supporting a human group remain of small extent.

In systems where livestock is not involved, the volume to be transported is far less, which should widen the radius of distribution. But as the removal must be done on people's backs, the distance between fields and dwellings must remain moderate, generally of the same order as where livestock plays an important part. But there are instances to the contrary, such as that of Yoruba towns: there the citizens have fields up to 30 kilometres away and have no hesitation in carrying the yams that form the major component of their diet over these distances.

As long as humans have at their disposal only their own energy and that of their animals (when they have any), the ecological pyramids they use are inscribed in circles with a radius not normally in excess of a few kilometres. Each forms a relatively enclosed unit, even if flows cross its boundaries: animals and certain valuable products are easily moved and can be sold far afield, even when mass-produced items must be used on the spot. The ecosystems used by human beings are thus structured in blocks of limited size, on the scale of the farm and its surrounding lands in areas of scattered dwellings, or the village and its district in the open fields. Towns can take only a slight fraction of what is cultivated by these basic units and are unable to have it come from far: for essential products, not more than around 30 kilometres.

In the traditional world, before the double revolution of the mastery of fossil fuels and transport methods, humanized ecosystems thus appeared as units of small radius that were relatively autonomous. Within each of those cells, the vertical flows remained essential since the production of vegetable matter depended on them, but what people took for their own needs and those of the animals they bred was concentrated, apart from a limited horizontal movement, on where they lived. Some surpluses often appeared and could be used to feed a small town a few dozen kilometres distant. The area from which it drew its provisions thus constituted an ecological ensemble at a higher level.

Groups with a distant ecological support: horizontal relationships

The modernization which began with the revolution in industry and transport had considerable consequences for the structure of humanized ecosystems. Henceforth, foodstuffs could be dispatched over distances without the need to spend in doing so part of what the land had produced. The cost of transport went down rapidly and continued to decrease in the long term, so that the development which commenced around 1830 has still not ended: in the 1870s, cereals harvested in Australia arrived on the European markets at competitive prices. The ties that had of necessity existed until then between human groups and plant formations, which, in the last resort, would ensure that they were fed, became looser. In countries that import their grain supplies, demand is made on Canada, the United States, Argentina or Australia, depending on prices. The country called upon may vary from one week to the next. People drink coffee made of beans harvested in Colombia, Brazil, the Ivory Coast, Kenya or the heights of New Guinea. Products that have to travel long distances have less and less value per tonne: today people eat pork raised in Britain or the Netherlands on manioc flour from Thailand and South-East Asia. Thirty years ago, it was unimaginable that such a product could be the object of international traffic on such a scale (which explains why there is no protection for it within the European Community).

As regards production techniques, development did not take place in one fell swoop. During the nineteenth century, working techniques hardly advanced at all, and the growth in productivity resulted more from progress achieved in the rotation of crops and the wiser use of animal manure than from recourse to fertilizers or selection of species cultivated for improved breeds. Labour was plentiful: mechanization had not started and many workers were needed for milking animals. Everything changed with the invention of the tractor. There was no longer a need for animals to plough the fields or bring in the harvest. Modernization accelerated. Agriculture became so efficient that it no longer employed more than a small percentage of the active population in industrialized countries.

The transport of products provided by farming and livestock breeding was carried out so easily that the ecological supports on which human groups lived ceased to be formed by adjacent, continuous and stable areas. At least part of the basic foodstuffs of the large urban concentrations came from a great distance, and from regions that constantly fluctuated to the dictates of the market. Horizontal movements took on such magnitude that the relationships of societies with areas changed in nature. Groups no longer depended for their food on a portion of the earth's surface that was easy to pinpoint, known to everyone, and which people realized required every possible attention if it were to remain productive for a long time and regularly. The awareness of the ecological responsibilities that must be shouldered by human beings the moment they alter the circulation systems essential to recycling fertilizing elements disappeared.

With transport facilities and mastery of concentrated forms of energy, the problem of making restitution changed its nature. It is increasingly difficult to return to the agricultural areas whence they originated the fertilizing principles contained in the products of farming or livestock-rearing: on a global scale, no market exists for waste. The imbalances that tended to be set up in the small cells of the traditional worlds between areas close to dwellings and their periphery became accentuated. Resorting to chemical fertilizers luckily allows the problem to be resolved, at least quantitatively, in the zones from which, together with the harvested crops, some of the elements essential for the pursuit of crop-growing also depart. But other difficulties arise – for example, those relating to the increase in nitrate and phosphate content in the water-tables and surface water.

The problem of waste and effluents

The situation is more difficult in large towns where consumption is mounting: we do not know what to do with organic waste, which is burnt or buried in gigantic tips. To food consumption is added that which comes from the increasingly artificial environment with which humans surround themselves, and the use of energy which that implies. In place of the 10 or 15 litres of water used per person per day, now 200, 500, even 1,000 litres must be treated, which had been previously supplied to the consumer. The engines that use fossil fuels emit an abundance of carbon dioxide, carbon monoxide and complex products with a sulphur, nitrogen or chlorine base. In zones of industrial livestock-breeding, the waste from hundreds of thousands of animals must be got rid of within perimeters that are often narrow.

In natural ecosystems, for each cell there exists a limit to the quantity of fertilizers that can be transformed into living matter. Local surroundings are not capable of recycling enormous organic waste products spawned by the modern town. Solid or liquid waste is thus transported beyond the town's limits. Mineral matter is thus concentrated in the discharge. The ideal is to re-treat the rest and re-inject it into the endless round of natural circulation, but the quantities produced are so great that the ground where they are spread and the watercourse in which they are discharged are often unable to absorb them, and there is an increase in pathological imbalances.

It is therefore next to the flow of waste and refuse thus caused that humans remain linked to the ecosystems of the areas in which they gather. The problems caused by gaseous effluents are the hardest to solve. As long as atmospheric circulation is active, the wind carries everything away and dusts are deposited at a distance, sometimes hundreds of kilometres; but toxic gases act on plant life there too, and rain becomes acidified. In contrast, in calm weather the ecological thrust becomes local and living conditions become hostile to plants, animals and human beings.

Elements of ecological structuring of humanized areas do not disappear with modern domestication of concentrated forms of energy, but change in form and scale. The ease with which food products can be transported has freed groups from being obliged to live close to the nearest rural areas, but the elimination of waste and its reintegration into ecological pyramids continue to be carried out in relative proximity to inhabited places, whatever their size: up to a certain distance from the heart of the towns, this introduces new dependencies between neighbouring vertical ecological cells.

Further Reading

Binns, T. (ed.), *People and Environment in Africa*, London: Wiley, 1995, 286 pp.

Goudie, A., *The Nature of the Environment*, Oxford: Blackwell, 3rd edn, 1993, 416 pp.

—— and Vines, H., *The Earth Transformed: An Introduction to Human Impacts on the Environment*, Oxford: Blackwell, 1997, 256 pp.

Johnston, R. J., *Nature, State and the Economy: A Political Economy of the Environment*, London: Wiley, 2nd edn, 1996, 288 pp.

Owen, L. and Unwin, T., *Environmental Management: Readings and Case Studies*, Oxford: Blackwell, 1997, 448 pp.

Simmons, I. G., *Changing the Face of the Earth*, Oxford: Blackwell, 2nd edn, 1996, 464 pp.

——, *Environmental History: A Concise Introduction*, Oxford: Blackwell, 1993, 224 pp.

——, *Humanity and Environment: A Cultural Ecology*, London: Longman, 1997, 328 pp.

5

The Economic Foundations of Regional Geography

5.1 Society, Economy and Space

Geographers seek to understand societies as living realities. Societies are not organized in abstract space, detached from material contingencies, and then projected onto the earth's surface with a few adaptations due to natural constraints. They do not inscribe in a platonic universe ideas that are imperfectly reflected in our world. They are made up of people who live at a given time in a given place, and from the space that they possess, control, develop and build an infrastructure; everything that is accumulated there constitutes a heritage of use to everyone even if it is unevenly shared. The individuals that constitute societies communicate by codes that are learned in infancy or assimilated later. They share practices, react to symbols and develop systems of representation that permit them to understand their environment and to make sense of the world surrounding them, their life and the community of which they feel part. They construct in common, conceptions of the world and of society that lead them to define norms. The list of functions assumed by societies is not determined by nature but results from their activity, their inventiveness and the richness of the heritage that they draw on: it varies from one place to another and changes with history. All groups, in common, have however resisted the trials of time. They have suffered catastrophes, experienced wars and suffered scarcity and famines but, at the worst moments, their food supply remained sufficient to assure collective survival. At times of stress, the birth rate did not totally collapse, or normalized relatively quickly to ensure the replacement of generations at least partially. The culture permits adults to hold the expected roles for which they were prepared during their youth. A minimum analytical framework can thus be applied to all societies: it stresses their aptitude in all circumstances to produce a certain quantity of food and to assure order, without which family groups could scarcely protect and educate their children; it includes also the systems of communications that structure the groups.

Collective life functions in a concrete frame. Space provides energy, foodstuffs and the raw materials indispensable to the group and serves as a support to all activities of processing and services. It constitutes the framework of everyone's existence. It is therefore, according to the location, a factor of production or the object of consumption. As such, space is unequally evaluated: some places are agreeable, restful and fill existence with charm. Forms of interaction implied by social life require a setting in which to take place. Whether it is for the use of the individual or the collective, space takes on values: it is not simply a support. It is part of the universe of signs that humans always leave behind.

Remoteness acts as a break on the mobility of people, goods and information, but along equipped routes the obstacle is reduced: the world is thus traversed by routeways, which permit distance to be overcome at least cost and in minimum time; the world ceases to be composed of undifferentiated surfaces and becomes structured as a function of networks assuring flows in the most efficient conditions.

The roles that space plays in the life of humans are diverse. Generally they involve precautions, developments or infrastructures: crops must be protected by enclosures from the jaws of livestock herds; fields must be cleared of stones, levelled, fertilized, irrigated, etc. Journeys are difficult unless roads have been developed or railways built. All this implies a differentiation of land uses and the definition of individual and collective rights of usage, passage, of improvement, etc. A society cannot be understood if one disregards its basis of land ownership.

To analyse spatial aspects of group life, it is convenient first of all to isolate economic factors. It forms a link with ecological aspects, which were proposed in the previous chapter: food products, raw materials and energy are taken from the environment, which introduces a tendency towards a dispersal of social life. The necessity for relations, on the other hand, encourages concentration.

In our world, a good proportion of manufactured goods is destined for distant buyers, but the freedom of consumers and producers to locate where they wish is not total: remoteness imposes difficulties of communication and the movement of people and goods takes time and requires specialized infrastructure and equipment. Distance has a cost, which varies with the technical level and environmental conditions; it is much more difficult to maintain roads in mountains or regions of extreme climate. The Siberian cold or the humid heat of the tropics destroy the surfaces. Since the nineteenth century, modern means of transport have greatly lowered costs; they have opened up the world to massive trade. Areas unserved by ports, railways and good roads are marginalized.

To understand how space is divided and organized, it is necessary to dwell on the impact of transport and communications on social life. Spatial economics, which has been progressively constituted since the beginning of the nineteenth century, has elaborated concepts and methods in this field with which the geographer can usefully familiarize himself and master.

5.2 The Organization of Economic Space: Natural Aptitudes, Distance and Homogeneous Regions

The Conditions of economic life

Humanity has multiple needs. Nature does not usually supply what we wish where we need it. What we consume must be produced by mobilizing knowledge and expertise, using the energy supplied by workers, added to the contribution of animals and engines and using tools or machines. When the resources used do not originate in the same area and when production takes place far from the customers, the entrepreneur must know where to procure resources and where the demand he seeks to meet is located.

The economic world is characterized by the unequal distribution of primary production of agricultural foodstuffs, of minerals or energy, by the very contrasted distribution of production techniques and market knowledge and by differences in levels of equipment. The economic world is also characterized by divergences in the quality of life from place to place. Humans seek to obtain the best possible existence. This explains the demand for goods, services (that is benefits furnished directly by other people and which generally incorporates a high proportion of information), and pleasant environments. To achieve this we must become producers. People are attentive to natural endowments, infrastructure and facilities, technical and commercial skills and all the information relating to market conditions.

People have only a limited time available; the techniques and skills that each person can master cannot be highly diversified; the natural resources that can be obtained from a given place are variable, even in a rich country they are always limited: climate only suits certain crops and the subsoil only contains a few useful minerals. Economic life suffers severe limitations when it takes place in a narrow local setting. A Robinson Crusoe cannot feed himself in a sufficient and regular manner, dress comfortably, house himself decently or have access to necessary tools or services necessary to his education, health and recreation if he is reduced to his own capabilities and cannot count on the possibilities offered by his island. An economy only grows when exchanges develop. This implies transport of goods, movement of people and transfers of information. For societies that are technologically poorly equipped, the movement of objects or people is so expensive that the majority of economic relations are locked into their locality. When the cost of mobility is reduced, circuits are expanded but unequally. Certain relations continue to be enacted at the local scale while others embrace the region, nation or the world scene.

Self-sufficiency and natural conditions

Self-sufficiency prevails as long as the means of transport and technology are ineffective. Each group struggles to produce, within a limited radius of its settlement, food, clothing and the tools it needs. The division of labour is limited. Weaving, wickerwork and wood-working are widely distributed. Fire-using techniques – pottery, metallurgy – are often the preserve of specialized castes. The spatial framework is made up of small-scale juxtaposed cells. The crops grown vary often from one point to another: the problem of self-sufficiency cannot everywhere be resolved in an identical manner. Human food needs remain more or less the same whatever the environment but there are differences in the natural aptitudes. There are environments where cereal yields are heavy and the pasture rich. These fertile lands are contrasted with others which are less well-endowed. Instead of white bread, one ate rye or mixed wheat and rye, oats or barley, porridge, buckwheat cakes, tortillas or maize polenta. Elsewhere it was sweet chestnut that provided the majority of starch. Since the eighteenth century, the life of many European countrysides was transformed by the potato. Human groups had to invent ways of life capable of extracting from the specific setting in which they lived what was necessary for their food supply.

These ways of life are the product of practices and received experience repeated from generation to generation. They are structured around the combination for food of crops, livestock, gathering and hunting to ensure the survival of the group. To that is added a variety of skills: spinning and weaving, the treatment of skins and furs, building, wood and metal-working, based on local raw materials.

These systems, and the landscapes and developments related to them, occupy more or less large areas; at their heart, the same heritage of experience is used. Beyond that, often quite abruptly, another type of habitat dominates due to a different combination of plants, dietary and clothing habits. Societies based on a specific way of life are still very much in evidence in all parts of the world which have remained marginal to the industrialized world – in Africa or above all in the mountainous regions of Asia. They have also left signs that persist in the agricultural landscape in the old homelands of Europe and have left their traces in alimentary regimes and mentalities.

Do the boundaries of ways of life correspond to different environments? Yes, to a certain extent, for ecological reasons: one could not live on wheat bread in an acidic soil environment before the transport revolution. But the temptation to make a determinist interpretation of these distributions must be treated with caution: many of the characteristics of traditional societies are not linked to their environment. It is in the direction of the diffusion of techniques, the unequal distribution of knowledge and the greater or lesser resistance to innovation that an explanation must be looked for.

The obstacle of distance and the notion of threshold

Humans need products to satisfy food needs, clothing, housing, provision of tools or mechanical aids and the services which distribute products, facilitate movements and journeys, assure education and health and maintain order internally and externally. The economic system permits the putting in place of all the factors necessary for the activity of productive enterprises and to transport the products obtained towards the distribution system. Products are thus made available to households, which acquire and consume them. To the transport of necessities must be superimposed the transfer of information, which permits the process (one can only produce if one knows that a demand exists), the movement of people along with the goods and the monetary flows that transact the whole operation.

The economic circuit is set within spaces of varying dimensions as a consequence of the unequal mobility of goods and people that it activates. All movement implies costs; the physical movement of goods and people demands infrastructures, vehicles and energy; in the case of people there must be added cost of time and fatigue. This explains why the transport revolutions have not expanded markets for goods equally (often at a world scale today) and the distances that people are prepared to travel to acquire services (which, according to the case, may be in the order of a kilometre, ten or a hundred kilometres, rarely more). For each movement there exists a relationship between the maximum amplitude that could take place and the advantage expected. Economists express this by stating that there exist spatial accounts: a product that is located immediately adjacent to a consumer has more value to him than one which requires him first of all to make a journey. The amount that he is prepared to spend to acquire it diminishes with distance. Beyond a certain distance, the item has no use or value; it has a threshold limit, often high for a material item, very restricted in the case of a service. This creates circles around points where services are offered within which they are genuinely accessible. All the points located within this circle of threshold limit have in common the ability to be served by the same centre.

The principle of threshold limits structures the life of inter-relations: daily movements on foot should not take too much time nor be too tiring – a few hundred metres walking. Periodic journeys, recurring every few weeks for example take place over a wider radius, but still remain moderate. The networks of interaction are thus circumscribed within circles with variable dimensions as a function of the frequency of movement. Daily movements take place over a limited space: that of the apartment, the house and the garden where the household lives, that of the street or district where the children play, go to school, and, until relatively recently at least, one purchased fresh foodstuffs. Other commodities or services can be acquired further away. The circles of professional relationship, and those which develop between producers and the distribution system, are enlarged to national dimensions or further. Since the research of Walter Christaller and August Lösch in the 1930s, economists and geographers have become very aware of the influence of threshold limits and the role of central places in the structuring of space.

The life of inter-relationships thus results in a division of space into areas which are integrated as interaction fields. This is valid at all scales and especially at those where economic factors apply on a daily basis.

The role of localities

At the level of households, the function of the economic circuit implies:

1 the to and fro of workers towards the field, the workshop, the shop or the office;
2 movements to buy various goods and the multiple services that everyone requires.

The movements linked to employment and consumption take place within areas of modest dimension: people only take a few tens of minutes per day to reach their workplace, to do their shopping and to meet the partners that they interact with most often. On foot, half an hour is already substantial. It is exceptional for one to have the time and energy for a daily return journey on foot of more than an hour.

The circles enclosing the trajectories traversed by members of groups in their daily life have radii that, generally, do not exceed 2 or 3 kilometres. The fabric on which traditional society is built thus is very fine grained; it corresponds to that of the commune, which still constitutes the basis of political and administrative life of France. Their average area of 14 square kilometres corresponds to a circle of a 2.2 kilometre radius.

When one has access to public transport or cars, one can traverse in several tens of minutes 10, 20, 30 kilometres or more. The perimeters of daily life change their dimension: the localities assume new forms and extend considerably when access is restricted to train or tramway, the enlargement of the areas is very uneven; one is enclosed within the 2 or 3 kilometres accessible on foot in most dimensions, but one passes to radii 10 or 20 times greater along public transport lines. As the routes diverge from a centre, the interaction field extends as fingers in these directions: at the end of the nineteenth century this seemed so remarkable that the theme of the tentacular town became an obsession.

With the motor car, mobility again became isotropic. Urban places resumed their previous, roughly circular, shape, but with much larger radii. The image of a sheet spreading out indefinitely captured the imagination.

Within the lowest-level circles, partners meet often and for a variety of reasons. Information circulates easily without involving any costs in particular. Many aspects of social life are facilitated; economists state that certain benefits of external economies result. This gives rise to social ensembles within a space essentially structured by the effects of interaction fields.

The logic of the location of productive activities and the formation of specialized regions

Since movements are costly and information about distant conditions must be paid for, the distance from sources of supply and from markets influences the profitability of activities. Producers locate where they can gain the best results. If they are unable to relocate, they select the most profitable speculations in the place where they are sited.

An enterprise exploits the technical abilities provided by the labour force, the technicians and engineers that it employs, or by the consultants it engages. Some of this know-how is incorporated into the tools and machines employed. Semi-finished products and components acquired from elsewhere also permit an enterprise to complete its own skills by those obtained from other firms. The skills and aptitudes needed by productive activity are slow to be acquired: they are therefore concentrated in the places where they were assimilated. Highly qualified engineers or technicians can be attracted from elsewhere, but this depends on offering advantageous conditions: the stock of technical expertise is normally confined to certain zones. It is possible to transfer it elsewhere, but the cost to be paid is high.

Production has the objective of supplying raw materials and transforming them. To achieve this, the process of production mobilizes, as well as high skills, energy from a variety of sources. Energy is supplied by the sun in the case of agricultural or forestry production, by animals or by people where it is the physical strength of the workers rather than their intellectual aptitudes that is needed, or by motors which transform natural forces contained in fossil fuels or in the nucleus of certain atoms into mechanical or chemical energy.

Productive plant is in part mobile and in part immobile. The land, (fundamental in agriculture), buildings and the majority of transport and communications infrastructure are non-transferable. Tools and machines are transferable: 'turn key' factories can be sent in component form to the other side of the world. But this cannot take place without very high cost and much risk. Location factors thus vary according to the extent to which immobility is more or less significant.

The entrepreneurs guide their most favourable location choice by an analysis of global costs. These vary from one place to another: one could say that they form a surface in which the lowest point is the most favourable. Economic logic thus requires that producers in the same sector should all group together in the same place to achieve the highest profits (figure 5.1a).

The profit surface has various forms. It can be that it has such a tight maximum that concentration over very limited areas is imposed. This was the case for industries at the beginning of the industrial revolution. Transport was still very expensive. The steam engine, which was the basis of mechanized production, has a very weak cost efficiency, since it consumes enormous amounts of coal. The transport of combustible material weighed so heavily on global costs that it was essential to locate at the pithead: the owners attracted workers, technicians and engineers who had the necessary competence, housed them and provided necessary services for them. They elaborated transport and communication infrastructures. Thus, this resulted in the formation of totally specialized industrial regions: space was taken over by mines and factories and by the heavy transport infrastructure necessary to both, and by housing for its workers and employees.

Today, the tyranny of transport is weaker. The profit surface that it imposes portrays relatively extensive high zones within which the variations are sufficiently weak, so that the influence of other factors can multiply the number of competitive locations (figure 5.1b). In the case of industrial activities, the area taken up by factory installations and

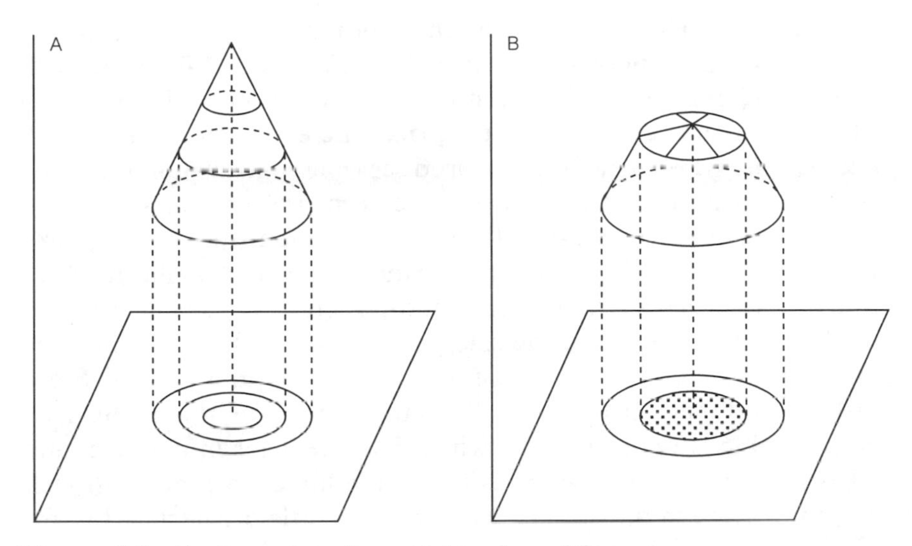

Figure 5.1 Surface of profits and location of firms
On the vertical axis, the level of firms' profits. In A, the profit surface appears as a steep-sided cone: the best location is a point. In B, the profit cone is truncated: the optimal locations appear as zones (indicated by dotted area).

the related infrastructures occupy a much more modest place in the landscape. We still term these as being industrial regions when employment in this sector is dominant.

When the relative importance of transport costs is reduced, entrepreneurs become sensitive to other cost aspects facing them. They seek places where the necessary skills cost less. For relatively unsophisticated manufacturing they take advantage of the improved education provided in Third World countries, whereas for those which require scarce skills they choose pleasant locations where it is easier to attract and retain highly-skilled labour. When acquisition of marketing information weighs heavily on the balance sheet, the entrepreneur seeks to locate in major economic centres which gather information from the entire world.

To the extent that the diverse operations undertaken by a firm find their optimum conditions in different locations, it is obliged to disperse its plants. It adopts this solution, however, only if the internal communications costs are not too high.

Very specific conditions apply in the case of the agricultural sector: production depends on the exploitation of a free but dispersed source of energy, the sun. Vast areas are required: fields, orchards, plantations, vines or meadows cover virtually the entire space in areas which are universally productive. The entrepreneur never locates in an empty space. His location is generally imposed by the strength of competition:

he settles where he finds land to purchase or rent at a price he can pay. As the substantial landowner, Johann Heinrich von Thünen demonstrated at the beginning of the last century, the problem is to choose what will bring the best returns at the place where the holding is located.

Let us analyse what a hectare produces on a given holding. The return varies with the crop selected. The crop is placed on a market – a town where consumers exist. Taking into account the quantity produced and the prevailing price, the farmer obtains an amount which is different for each particular commodity. To calculate the net revenue, he must deduct from the sale price obtained the costs incurred in transporting the produce to the town where he finds buyers. Since foodstuffs lend themselves unequally to transportation, those which are most profitable are not the same when distance is taken into account. Market garden produce, which is high in volume, fragile and travels badly, gains close to the market. Meat returns little per hectare but the herds can be led on foot to market producing very low transport costs: extensive animal rearing for butchering is suited to peripheral regions. Crops thus are distributed in concentric circles around markets.

In practice, the choice of crops is made by taking into account both the natural aptitudes of the environment and the distance from the market. This is what configures the American agricultural belts (figure 5.2): they become progressively more extensive as one goes away from the market, essentially situated in the east, towards the west, but the choice of crops also reflects a decreasing rainfall. From north to south, the contrast results essentially from the nature of the soils and rhythms of climates.

Economic logic leads therefore, when trade becomes possible, to the formation of specialized space. Alongside agricultural regions, industrial regions are defined.

The advantage of central areas and the formation of complex economic regions

When a relatively extensive economic space is closed, all its parts are not equally attractive. Activities that have to support high costs to market or for contact with customers have an interest in locating in central areas. To evaluate this advantage, measures of population potential may be employed.

In 1947, the astronomer J. Q. Stewart took as a starting point Newton's law of gravity to model the strength of interaction manifested between two populations living within a territory: the attraction exerted by a point on the whole of the other points varies as a function of the population living there and is inversely proportional to the squares of their distance. Take a set n of places with a known population. The

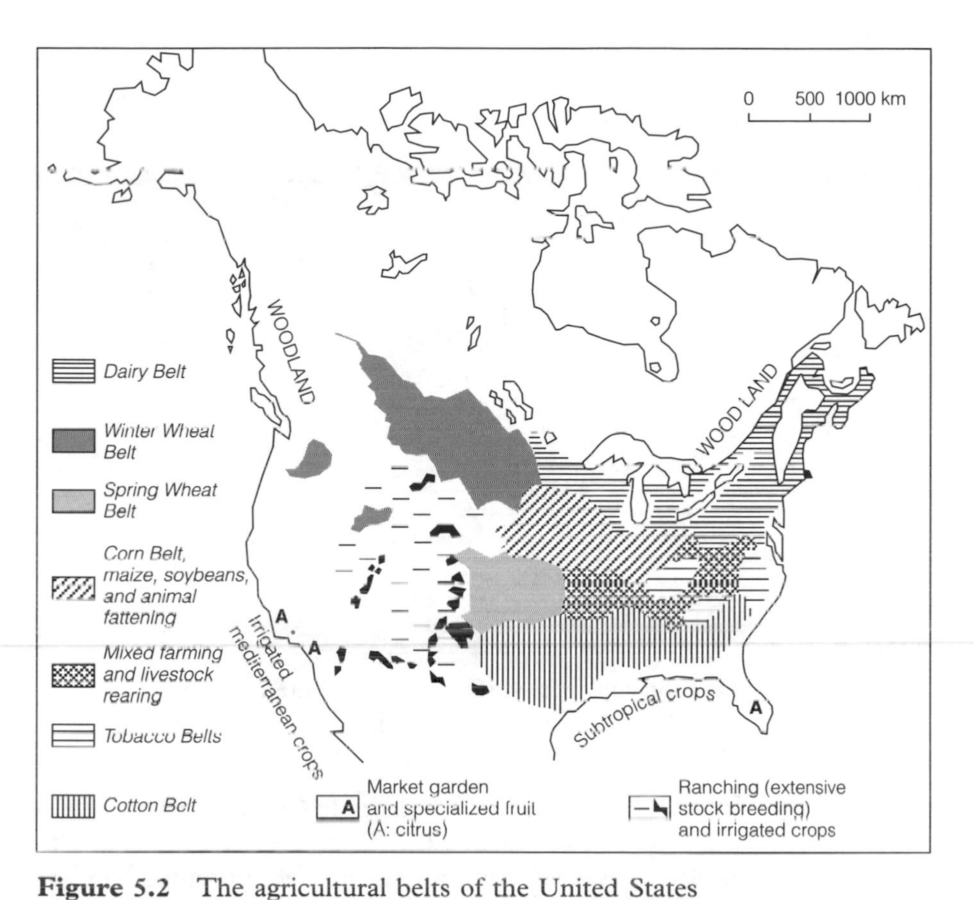

Figure 5.2 The agricultural belts of the United States
Source: P. Claval, *La Conquête de l'espace américain*, Paris: Flammarion, 1989, p. 155.

population potential of any place *i*, with a population of P_i is equal to the sum of the influences exerted on *i* by the other places: the influence of *j* is for example the quotient P_j/d_{ij} of the population living in *j* at the distance between *i* and *j*. The formula which gives the total potential is written:

$$\text{Potential in } i = \sum_{j=1}^{n} P_j/d_{ij}$$

The influence exerted by the population P_i on itself will be infinity (the distance *i* to *i* is zero), which is absurd. It is for this reason that, conventionally, P_i is divided by the half distance to the nearest point.

We can then draw a map where the isolines join the points of equal potential. This permits a visualization of accessibility fields: the central zones are those with the highest potentials. They offer entrepreneurs relatively uniform transport costs. The majority of their points can receive industries and services attracted by the relative proximity

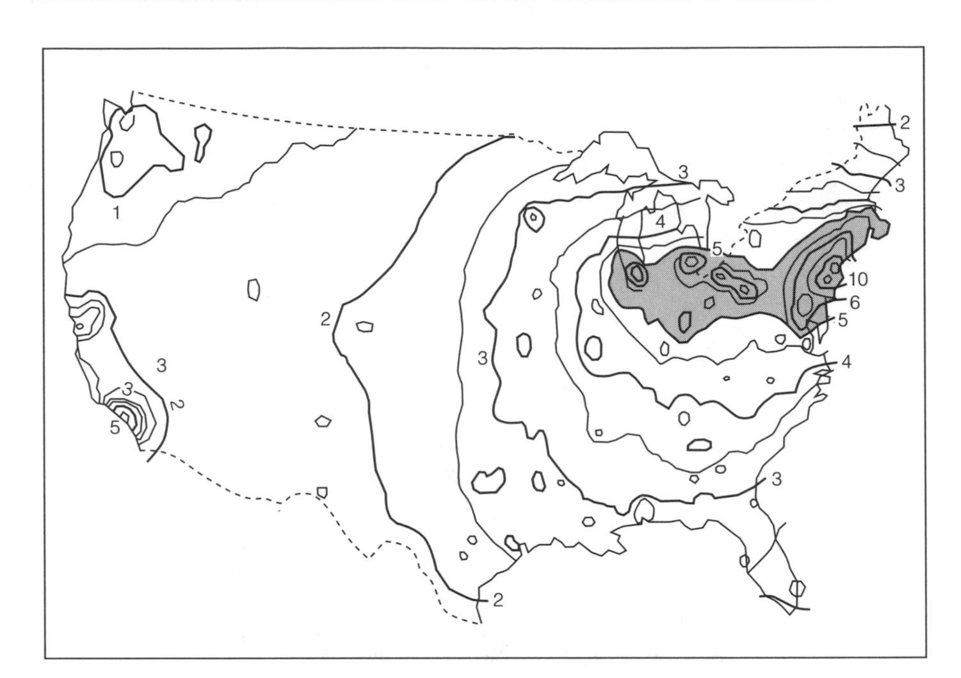

Figure 5.3 Population potential surfaces in the United States
Source: B. J. Berry and J. Kasarda, *Contemporary Urban Ecology*, New York: Macmillan, 1977, p. 274.

of the market. In the United States, for example (figure 5.3), the areas with high potential cover the North-East, from the Atlantic Coast to the Mississippi valley. Another zone with a high potential surface exists in California. The whole area constitutes the industrial belt, which combines manufacturing and service activities with a national or international range.

Central areas revealed by maps of potential derive, as with peripheral regions, from their natural aptitudes or from their technical and social heritage, and specialize in types of production where they enjoy the greatest advantage. But they also attract activities whose results depend directly on accessibility to the whole of the market. This is especially the case for light industries and certain categories of services. Their activities are thus very diverse. These are rare regions of complex economies. As the range of employment becomes more elaborate so new advantages appear. First of all, firms arrive, which are able to place certain manufacturing stages with subcontractors capable of achieving economies of scale by virtue of combining multiple orders. Also, information circulates more rapidly because the partners are more numerous and varied. Central areas thus benefit from externalities of scale and information. A study of the conditions in which goods, people and messages circulate helps to explain the origin of some of these advantages.

5.3 The Role of Networks in the Economic Organization of Space

Transport, movements and information flows follow specialized infrastructure: tracks, roads, railways, telecommunication lines, etc. In economic terms, the distances that count are not those measured in a straight line between two points (as we have done for the purposes of simplification in the previous sections) but those which follow equipped routes. Costs do not only depend on the distance in kilometres: they reflect the quality of the transport equipment and facilities, and depend on how the networks are structured.

The logic of transport

The cost of movements and of exchanges of information depends on the physical nature of the spaces to be traversed and on the way in which they are equipped. Humans do not live on an isotropic surface but in space structured by networks. Geographers have long underlined the irregularities introduced by the layout of relief and dense vegetation, and the interruptions caused by watercourses or marshland. They were aware that the disposition of rivers or the existence of gaps facilitated the convergence of routes at certain places. But choice came into being when the alignment of transport lines introduced even stronger irregularities.

Large-scale facilities (railways, motorways, deep-water ports) if exploited to a maximum capacity, permit the lowest transport costs per tonne/kilometre. To maximize economies of scale it is necessary to design the whole infrastructure systems with the aim of converging traffic over the best-equipped routes (figure 5.4). At the time when most movement of goods was dependent on pack animals, infrastructure had scarcely any influence on transport costs. The situation changed with the introduction of wheeled transport: good quality roads requiring less traction effort. Today, from the track to the road to the motorway, from single-tracked light rail to electrified lines capable of being used by high-speed heavy trains, the range of possibilities has become very wide; the hierarchization of well-equipped itineraries is a major consideration of economic life.

The logic of communications

In the case of communications (defined as contacts and exchanges of information between partners), the problem can be understood by

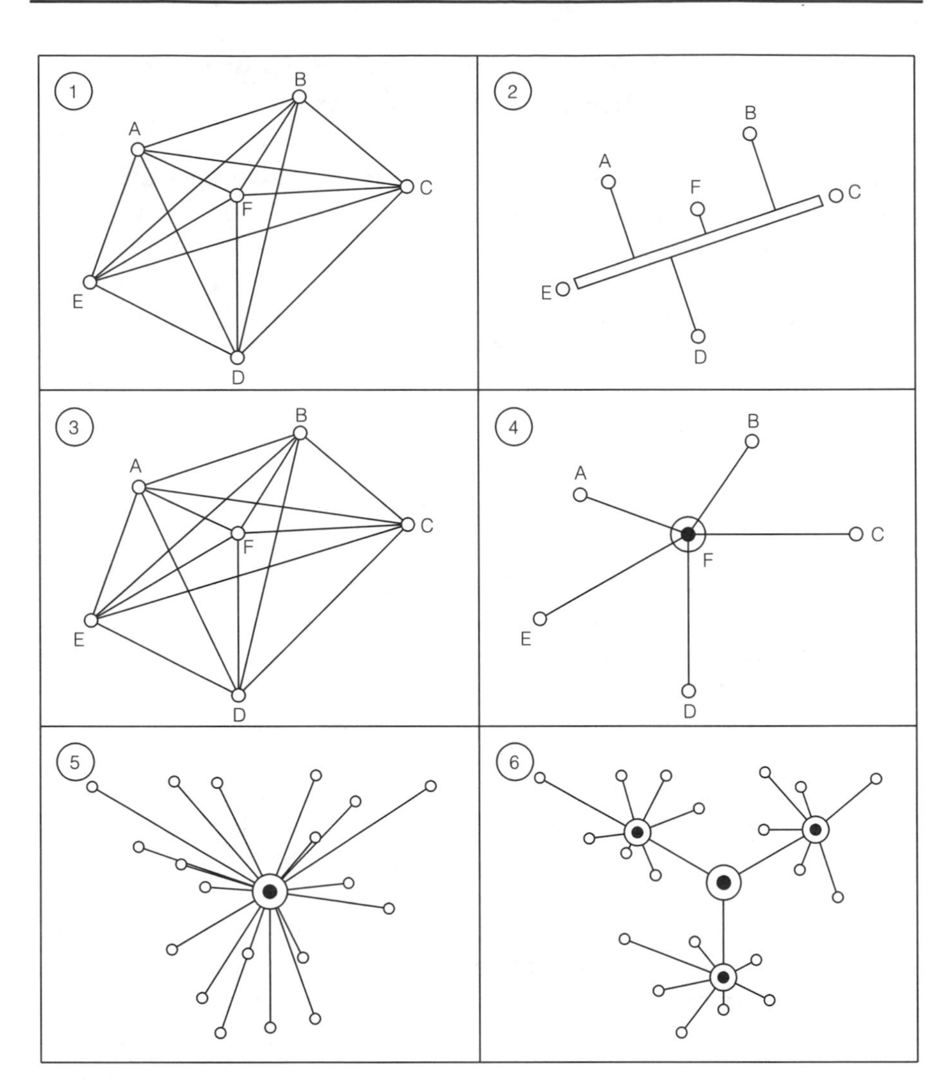

Figure 5.4 The logic of transport (1 and 2) and the logic of communications (3, 4, 5 and 6)

To facilitate freight traffic between the points A, B, C, D, E and F, it is not the line of direct linkage (1) which is the most advantageous, but the reduction of all the traffic (2) on to a well-equipped axis.

To facilitate the flow of information between A, B, C, D, E and F, it is not the line of direct linkage, which multiplies the length of lines and the number of exchanges (3), but the linkage of all with a central switchboard (4) that gives the least cost. When the partners are spread over a large expanse, the solution of a single switchboard (5) extends the lines uselessly. By creating local switches linked to a central switchboard (6) the total costs are minimized.

The transport economy leads to a hierarchy of lines, that of communications leads to a hierarchy of centres.

analogy with the installation of a telephone network. Each subscriber cannot be connected directly to all the others: this would require each person to house a giant switchboard and an astronomical number of lines (the formula states that this number N is equal to $n(n - 1)/2$. where n is the number of subscribers), and kilometres of cable as a result. A telephone exchange to be situated at the heart of the zone permits the minimizing of connection charges and the number of lines to be installed. The length of lines is reduced to a minimum if it is located at the centre of gravity of the subscribers.

The density of local calls is much higher than long-distance ones so the best solution is to establish a hierarchical network of exchanges that permits complete coverage of the area.

Direct face-to-face communication between partners was for a long time the only possible means. This still remains irreplaceable in spite of more and more sophisticated means of long-distance communication at our disposal. The church, the village square where people meet, are the local connections in peasant society, the café for men, the wash-house for women are the high spots for the exchange of news.

The small towns where periodic fairs take place serve as connections at a higher level for the people visiting them. This is the means by which a hierarchy of meeting places, of central places, permits members of an entire society to make contact.

The logic of combined networks

The logic of communication leads to a hierarchy of points; the pure logic of transport favours lines. To the extent that a good number of relations take the form of contacts, of face-to-face interviews, the communication is preceded and followed by movement (or two of the partners arrange a meeting at a point easily reached by both). The optimum form of networks results from the combined logic of transport and communication. Meeting places are located at junctions of the overall route network. They have a hierarchical structure and are ranked more highly in proportion to the size of the area that they serve. The lines that drain dispersed population towards the nodes, and which put these people in contact with one another, are also subject to a hierarchy: they are equipped with an infrastructure which varies as a function of the traffic they carry. The most elaborate and effective infrastructure is restricted to the axes carrying the most traffic.

Manual technical training, the acquisition of a service, the majority of commercial relations, the expression and settlement of conflicts, etc., all types of complex relations where the precision of the information, sentiment and good faith come into play, demand personal contacts. These are multiplying in our world. Life based on contact thus is articulated by networks established to permit this type of exchange: the axes are represented by roads, railways, motorways and airline routes; the nodes are constituted by towns the central areas of which, the business districts, are basically simply gigantic connectors permitting one interlocutor to pass to another in the minimum of time.

Relations between people extend quite widely over space but are not all circumscribed within the same circles, since they do not correspond to the same requirements and do not provide the same satisfaction. If networks were drawn directly from one individual to another, the world would be criss-crossed by an infinitely dense linkage of diffuse flows circulating in all directions and there would be as many nodes in these networks as there are families or minute settlements. Because of the advantages resulting from a well-articulated transport and communications network, relations are organized around poles and use a hierarchy of routes. Thus constituted, these are the networks which in reality organize space.

The hierarchical segmentation by which contact space is organized springs from the interplay of adjustments and from the composition of individual decision. It is in centres that certain categories of commerce succeed best and the businessmen practising them agglomerate there. Thus they strengthen the attraction of the agglomeration where they have located, and contribute to maintaining a high rank in the urban hierarchy and to the justification of telecommunication and transport systems which in turn offer advantages: the process is cumulative.

Polarized regions

The lowering of transport and communication costs is marked by a multiplication of all types of circulation: goods travel further and their markets become national or international. To supply factories and distribute finished goods, a greater volume and detail of information flow is necessary. The postal service and new methods of telecommunication, telegraph and telephone facilitate the transmission of information, but very often business demands contacts: it is indispensable to meet one's partners to appreciate risks, to understand the needs of clients and to judge the opportunities for new manufactures.

The space within which the modern economy develops is not structured uniquely into homogeneous fields, as the concept of major specialized regions tended to suggest. It is constructed of networks: transport networks which often transit goods and people and, increasingly, by specialized telecommunications networks. Technical progress confers economies of scale on networks: it favours the concentration of flows of goods and people on the best-equipped axes, and the creation of a hierarchy of information exchanges, the best situated of which take a growing proportion of long-distance exchanges. The towns, which function as support systems to the nodes in these two types of network, see their role increase. The networks that they define are transformed to reflect the transport and communications systems that give them life.

The incessant reinforcement of the importance of the higher ranks of information exchange is translated into a burgeoning of those towns that guarantee transparency over large areas: virtual metropolises thus structure economic life. They owe their prosperity to the gamut of intersecting flows, to their trading houses, their markets, their banks and also to their stock exchanges. Their newspapers ensure the same news service to the hinterland. From throughout their zone of influence, people come to consult their experts and to apply for loans when wishing to start new activities. It is there where conferences and exhibitions are organized and where fashions are born.

Modern economic space is thus organized by cities, which divide space into zones of influence. In the nineteenth century and at the start of the twentieth century, regions were increasingly structured around metropolises capable of assuming functions of meeting, exchange and animation. They are dominated by the national capital even though international flows are organized by the major ports. This type of spatial organization, especially well illustrated by certain regions of the central core of Europe, extending from the northern London basin to northern Italy, has sometimes been considered as a model. Where links are different in character and where cities are not well developed, state intervention often has the objective of correcting what appears to be an anomaly and a weakness; this is the way in which the French policy of creating *métropoles d'équilibre*,* launched at the beginning of the 1960s to offset the predominance of Paris and to achieve a more balanced demographic and economic distribution, should be interpreted.

The process of evolution continues and today favours the major economic metropolises with an international vocation. They are more important than regional metropolises which until recently focused the attention of geography: they total two or three, or even just one in countries the size of France or Italy. In the United States, where their evolution is stronger, there are scarcely twenty in an area sixteen times larger than France.

* The eight cities or groups of cities designated by the French government for expansion to counterbalance the excessive power of Paris.

5.4 Space as a Consumer Good

The differentiation of daily life space

After working hours comes the time devoted to relaxation, sport, recreation and rest, the time for family life, social occasions, meetings or shows that people indulge in in animated places. A large proportion of space is devoted to and developed for this end: housing is built along with gardens, sports grounds, parks, entertainment halls or streets and squares where people enjoy meeting.

All the various combinations of land use do not offer the same advantages to the consumer. The latter, when choosing his home seeks to reconcile several objectives; a household has no interest in locating too far away from the workplace of the adults and the children's schools; it must have access to a flat or a house big enough to live comfortably and with a garden or nearby playground if the children are infants. Adolescents and adults are more attracted to the animation of nightlife and the range of shows available in the town centre.

The best location is not the same for everyone. In a town, young couples with children are more willing to go to the periphery where more space may be found; the unmarried or households without children are more attracted to central locations. Urban space is thus differentiated as a function of proximity to work, the abundance of services and recreation available and the size and quality of housing. Land and building prices reflect these conditions.

Neighbourhood space is structured by the preferences of the consumers. To a large extent, as has just been shown, rational factors enter into the calculation. But collective enthusiasm and snobbishness are equally present: one locates in one neighbourhood rather than another because it has a better image. To indicate on one's visiting card that one lives in Neuilly is more flattering than an address in Aubervilliers.*

The basic cellular unit in the life of a region constituted by neighbourhoods are thus very largely shaped by the needs and tastes of the consumer.

Amenities, tourism and the organization of space

Generally it is impossible to find everything one wants in the same place. If one were condemned to be absolutely sedentary, life would

* Respectively fashionable and working-class districts of Paris.

always lack something. For two centuries, our societies have appreciated how to offer their members possibilities to escape for several days or weeks each year and at retirement age.

Among the forms of economic specialization that characterize modern space, certain forms spring from the need to satisfy consumer need for open-air life, entertainment, atmosphere and culture. People leave grey towns where their work condemns them to live eleven months of the year, for the sea, the mountains or the countryside. At the time of retirement, it is tempting to move there and to flee the constraints that until then were tolerated without enthusiasm. These migrations reinforce the economic strength of spaces of pure consumption constituted by tourist regions.

The evolution has now gone further. The improvement in means of circulation gives an increasing number of activities great freedom in their choice of locations: their transport costs scarcely change from one point to another. Enterprises in these sectors therefore tend to satisfy the tastes of their managers and workforce, to locate in pleasant places. Without paying higher salaries they assure more tangible advantages on behalf of their employees. It is a way of retaining the workforce employed. High technology companies, which are always anxious to recruit a quality workforce, are very sensitive to this. They are the first to gravitate towards regions which have nothing to offer other than an attractive climate, their shores, their mountains or the historical legacy of their towns or countryside. The boom of California, the arid South-West and Florida in the United States can be explained in this way, as can the Midi and the Alps in France or Bavaria in Germany.

5.5 The Organization of Space and the Dynamics of Regional Economies

Economic life has created two forms of economic region on the surface of the earth: specialized zones (which are more complex, at the centre than on the periphery) and zones of urban influence. Their unequal dynamism is expressed through the displacement of their frontiers: for example, demand for a product may weaken and the zones where it was cultivated contract, nibbled away on their margins by speculations that become more profitable; or modern transport may bypass a town and the external sector of the area that it dominated turns towards other centres that were not left behind by progress. However, the influence of threshold-limits prevents the zones of influence at a given level from increasing indefinitely. This explains the relative stability of urban hierarchies and of the frontiers of the regions under their control. What changes is the total population and the volume of activities more than the contours of the areas of polarized regions.

The differential evolution of economic spaces is explained by various factors. Firstly, it reflects the structure of the global economy of a given space and the way in which demand has evolved. The areas specialized in production with falling demand tend to decline, while those oriented towards dynamic sectors have a better chance of experiencing prosperity. This is why the old coalfields, so dynamic in the past, are today losing ground.

But regional dynamism is not only controlled by what happens at the global level. It results from internal multiplier effects in each territorial unit. Part of the income generated there is also spent there and nourishes local activity. The more the economy of a territory is complex, the higher the proportion of purchases made locally. The multiplier effect of external relations is thus more apparent there: the opportunities for growth induced by the rest of the world are higher. For this reason, areas with complex economies (the central zone of developed nations generally), normally experience a stronger, more rapid growth than regions where productive activity is limited to a few sectors only. The multiplier mechanism, linked to the diversity of the economic structure in place, is one of the reasons for the maintenance and amplification of inequalities in development.

For the growth potential thus offered to be exploited, there must also be enterprising investors to seize the opportunities. In the main this is a question of information and transparency. It is essential to be informed of what is going on so as not to miss opportunities and to assess the risks accurately.

In the modern world, the savings intended for investment are gathered by banks and by other institutions in the financial sector. This facilitates a choice by individuals with money to invest. The concentration of available finance permits higher levels of investment than in traditional systems. But local opportunities in regions are often underestimated by the headquarters of banks located in the major centres. During a good part of the nineteenth century, this process acted in favour of the regional metropolises where the network of banks was established: as a result they experienced rapid growth. From 1880, 1890 or the 1930s, depending on the country, the centralization of financial circuits progressed rapidly; it was especially a redistribution of savings towards the national capital. The nationalization of the banking network helped the interventionist action to balance growth in a country like France, but control remained in Paris and the perception of the provinces only changed by stages.

Not all the money invested derived from national or international financial institutions. Entrepreneurs invested a proportion of their profits directly in their businesses. This is what gives regions that are the headquarters of numerous companies opportunities that are lacking in others. Small firms are often more active in this sense than large companies; they are willing to take more risks.

For a generation, the theme of bottom-up development, using the contribution of small companies, has become central to research on regional development in developed countries: in an atmosphere of cut-throat competition from newly industrialized countries, major companies tend rather to slant their production towards countries with the cheapest labour force. One cannot count on them to create industrial employment. The only thing that they do is to multiply their offices in cities well connected to world commerce. Thus it is local resources rather than the money markets that should be counted on to achieve regional development of employment. It is to meet this need that in France, regional stock exchanges have recently been revived by the opening of a second share market: it offers small well-known firms on the regional scene direct access to private capital.

In Third World countries, the risks run by investors, whether real or exaggerated because of weak transparency, paralyse enterprise. Only the resources capable of massive supplies at low prices on the world market for food, raw materials or energy are taken into account. To develop a rubber plantation, organize the production and treatment of groundnuts, open a mining or petroleum field, does not present major difficulties or risks and can be very profitable when the flow of production takes place without problems at a good price. The profits of the operation can be shared between the producing company and the local power, the guarantee of the relative security of the operation. They are rarely or very imperfectly reinvested. Confusing national income and private wealth, the local potentate often places his share in the protection of a western bank.

Investments to permit the growth of food or industrial production for the local market are rare: the inadequacy of transport and communication infrastructure, the gaps in manpower training and the doubtful reliability of local commercial circuits paralyse initiatives in many countries.

The very considerable gap in levels of education and wealth that were observable, and can still be seen, in many developing countries, also constitutes a brake. Dual social systems are scarcely encouraging for entrepreneurs: the rich elite prefers imported products because they can afford them, but also through snobbery. The standard of living of the great majority of the population is so low that few multiplier effects related to increases in wages can be expected. The stage when demand will be sufficiently abundant and varied to provide a market to stimulate local enterprises seems still to be a long way away. Chronic political instability, the unreliability of contracts, and institutional debt are also disincentives for future private entrepreneurs.

The situation is in the process of changing: the gradual opening up of developed economies has allowed newly industrialized countries, especially in Asia, to count on an ever widening range of export products. More advanced education and new possibilities for technology transfer have reduced the handicaps to essential skills for manufacturing and the running of companies. Local elites have emerged in many regions and are aware of their vested interest in diversifying activities and promoting modern forms of production.

Conclusion

The regional geographer has always accorded an important place to economic factors to explain the organization of space. The distribution of populations, their activities and their works on the earth's surface depend on what they are capable of extracting from the environment and the way in which they transform what they obtain and distribute to satisfy demand. When economic logic is allowed to operate in liberal regimes, adjustment mechanisms of decision-making and markets in particular, tend to institute a spatial order without recourse to planning or centralized management. Local areas find themselves structured by the activity of the local land market, which permits the distribution of land use to the benefit and preferences of producers and consumers. Above them specialized zones and polarized spaces are defined, which play an essential role in the regional structuring and formation of hierarchies of the world. The responsibility of the state is to oversee a sound organization of markets and to take corrective action to achieve an equitable distribution of incomes. Socialist systems, which purported, by absolute centralization of economic decisions, to impose the most rational solutions, have collapsed through failing to measure the importance of information circuits in the process of regulation.

The ecological and economic factors which we have reviewed only exceptionally reveal the most profound causes of the differentiation of space. We must turn towards social and cultural realities to perceive, underneath the apparent uniformity induced by the modern economy, the depth and permanence of diversity.

Further Reading

Amin, A. and Robins, K., The re-emergence of regional economics: the mythical geography of flexible accumulation, *Environment and Planning D, Society and Space*, vol. 8, 1990, pp. 7–34.

Boudeville, J.-R., *Problems of Regional Economic Planning*, Edinburgh: Edinburgh University Press, 1966, 192 pp.
Chisholm, M., *Regions in Recession and Resurgence*, London: Unwin Hyman, 1990, 217 pp.
Dicken, P., *Global Shift: Internationalization of Economic Activity*, London: Harper and Row, 2nd edn, 1992, 492 pp.
—— and Lloyd, P., *Location in Space*, London: Harper and Row, 3rd edn, 1990, 431 pp.
Hoover, E. M., *An Introduction to Regional Economics*, New York: Knopf, 1971, 395 pp.
Hoyle, B. S. and Knowles, R. D., *Modern Transport Geography*, London: Wiley, 1992, 224 pp.
Isard, W., *An Introduction to Regional Science*, Englewood Cliffs, NJ: Prentice-Hall, 1975, 506 pp.
Johnston, R. J., Taylor, P. J. and Watts, M. J. (eds), *Geographies of Global Change: Remapping the World in the Late Twentieth Century*, Oxford: Blackwell, 1995, 356 pp.
Knox, P. and Agnew, J., *The Geography of the World Economy*, London: Arnold, 1989, 410 pp.
Massey, D., Questions of Locality, *Geography*, vol. 78, 1993, pp. 142–9.
Richardson, H. W., *Regional Growth Theory*, London: Macmillan, 1973, 264 pp.
——, *Regional Economics*, London: Weidenfeld and Nicolson, 1969, 457 pp.

6

The Social and Cultural Dimensions of Regional Geography

The functioning of an economy often leads to an unequal geographical distribution of professional categories and incomes, but these contrasts are generally not sufficiently strong that the social dimension assumes a dominant role in spatial organization. It heightens the colours in the pictures painted by exchanges rather than drawing another.

Society is not reduced merely to its economic activities. Individuals are involved in the panoply of organized relations and this deserves our attention. The relationships that each one maintains with other people are intensely variable in nature.

Societies are not machines. Their members communicate and share the same heritage and culture. This explains why spatial organization reflects the manner in which knowledge and skills are transmitted, diffused and learned.

6.1 Specialization, Polarization and Social Geography

Economy and social geography in the traditional world

The social geography of traditional societies is never uniform, even when dominated everywhere by agriculture. The way of life changes according to the setting; zones with owner-operatorship may be contrasted with those with share-cropping, with tenant farming or large-scale farms employing paid workers. Casual workers and smallholders sometimes complement their income by indulging in artisan activities: the wife spins, the husband weaves. Alongside those who live totally or partially from the land, may be counted full-time artisans, notaries and lawyers and sometimes a doctor and inactive landowners. The latter may visit their land at harvest time or may live there permanently. The dominance of peasantry excludes neither nuances nor variety.

The essential contrast is drawn between towns and countryside. This does not however create marked regional differences: each territorial unit is structured by similar networks of small towns and villages. Social differentiation exists, but, like economic differentiation, is limited: how could it be otherwise when at the heart of each *pays*, and based on local resources, the majority of the needs and services must be delivered to everyone?

Social geography and the effects of modernization

The social geography resulting from the opening up of space ignores the opposition of ways of life introduced by the qualitative adjustment in regional characteristics, but it is characterized by a diversification of activities and specialization in the workforce. In the traditional world, the essential theme was the distinction between town and country and with multiple nuances within each category – small independent landholders here, *latifundia* and agricultural proletariat there; the towns were oriented to artisan and trade activities in some regions; elsewhere, centres limited themselves to living off the rents derived from the neighbouring lands. From then on, what would characterize space was the concentration of processing activities in certain zones. The dispersed artisan activity in the countryside, local towns or urban centres was replaced by major concentrations of workers. The new elites, owners and technical staff, were divided between the manufacturing districts and the towns where their relationships with suppliers, bankers and purchasers were interlinked.

As the range of manufactured goods widened and industrial machinery became more diversified and as consumption extended to new articles and new services, so the variety of employment categories induced by industries became greater. The complexity of advice necessary for firms was accentuated, the multiplicity of businesses and the care given to the final market asserted themselves: this led to increasingly diversified and hierarchical social pyramids within the central area that these activities looked to locate in. The fabric of the peripheral areas was much less diversified. Mobility and affluence nevertheless gave birth to a new field of activity there: tourism. The zones where it prospers attract a large variety of consumer services. They are often situated where the environment has not been too spoiled by industrialization.

The industrialized world is characterized by an uneven spatial distribution of its social category components. The significance of agricultural workers declines: they only exceptionally form the majority of the regional population even if, at a localized scale, their proportion has sometimes increased in rural communities where there is no other economic base than agriculture; increased mobility has caused the disappearance of local services, which have become more concentrated in the nearest centres or towns. The dominant social characteristic of the nineteenth century was the formation of gigantic working-class concentrations. Artisans were not lacking in traditional societies, but it is scarcely worth discussing them since they were dispersed amongst the peasantry or small towns. With the industrial revolution, workmen became a majority in the manufacturing zones. At times they represent 70 per cent of the workforce in communities devoted to heavy or processing industry. Their proportion is smaller in cities, where services are more highly represented, but they can sometimes reach 60 per cent.

In the last thirty years, the workforce in the secondary sector has diminished rapidly in the most developed countries whilst the tertiary sector has swollen. But the latter is very diverse to the extent that the social composition of cities seems less homogeneous than a generation ago. In terms of revenue, the evolution is marked by an increased proportion in the middle-income categories. Their level and types of consumption are similar but they do not see themselves as a solid social group.

The geographical distribution of social categories has not ceased to evolve. In the traditional world, contrasts from region to region were weak. Within each it was in the contrast between the country-dwellers and the townsfolk who often dominated them, that the most significant tensions were detectable. The industrial revolution opened up gaps and gave a strong working-class coloration, with intense class struggles, to the regions where manufacturing was concentrated. In the post-industrial world that we are entering, the working-class masses are dissolving. The group of professions that gather, process and diffuse information is now in the majority, but their vested interests are so varied that they do not react en bloc and do not have class feelings. For the mass of middle-income earners, the real conflicts no longer concern work: they are concerned with quality of life, protection of the environment and the problems of women's rights, youth and the elderly.

The mobility enjoyed by these people always leads, within the enlarged localities, to a more pronounced differentiation of residential neighbourhoods: the competition for pleasant areas or for those that confer prestige results in a concentration of the elites – the wealthy, artists, intellectuals – in certain districts, and of the marginalized, the poor and recently immigrated ethnic minorities in the most underprivileged and congested zones. The social problems of the present day world are to be found at the scale of the large agglomerations rather than the region.

6.2 Systems of Institutionalized Relations

Consideration of forms of activity and income yield only a poor representation of social reality. To have a more accurate idea, it is necessary to analyse the relationships which bind people together and dwell on those that, most frequently and importantly, are institutionalized.

Social relations are very diverse and are situated between two extremes:

(1) Some linkages occur within the family group. Intense and warm feelings and intimacy play an important role. They are characterized by confident and security-inducing contacts, but which are sometimes heavy to bear for the individual who cannot escape everyone's gaze and cannot escape from the control thus exercised over him.

(2) At the other extreme, it is the case of the anonymous individual in a crowd, paradoxically alone and weak, but free.

Space economy does not include just these extreme situations: it considers households' responsibility for consumption decisions but scarcely as a major topic, it is below the scale that is considered significant. In the production cycle, it is the individuals and firms that come into play, but their social dimensions and their organizational problems were late to be considered. Regional analysis postulates a more nuanced view of the construction of groups.

The diversity of systems of relationships

Relations between people are set within institutionalized systems, which constitute the framework of society and introduce a high degree of regularity into contacts. Pierre Gourou termed this 'techniques of social organization': he showed them at work in traditional societies in South-East Asia and Africa. Elsewhere reference is made to the 'social architecture' of these groups and of those characterizing modern societies. Techniques of incorporation condition the way of life of societies with little differentiation and control the distribution of laws and the division of roles in the urbanized and industrialized world. These systems of relationship are interiorized from infancy and constitute an ensemble of coded behaviours that clearly identify each culture. The corpus of rules, which institutionalizes social life, vary from one society to another. In order to organize itself each civilization disposes of a more or less rich grammar and vocabulary of roles.

The minimal language is that of kinship, which defines both the laws of descendence and those of marriage. Its variations are infinite as ethnological research has shown. Archaic societies also experience association (in particular through the age groups) and trade (by marketing, by centralized gathering and redistribution or barter).

Primitive groups are readily egalitarian, but generally they do not escape from the effects of influence, domination or power. These can breed widespread inequalities, as in societies based on orders or castes, clientism or sworn faith as in the feudal system or the imposition of force, as in tyrannies.

Social relationships develop better when the rules to be followed are accepted by the largest proportion of the population. Partners engage more readily in transactions when they feel that they respect the same values and have the same objectives. The solidity of social systems is reinforced by the adhesion to the same beliefs and the same ideologies. The installation of a legitimate authority thus seals the pact that facilitates collaboration between people.

State institutions begin to take on a modern aspect when concentration of power and authority occurs to the benefit of one person. But real control over a population and the implementation of policies, involves the influence of a new type of organization: they are built around a particular purpose with staff recruited for this purpose which cement common interests and share an ideology. It is with the birth of administrative bureaucracies and of companies, which constitute the equivalent in the private sector, that the arsenal of techniques of incorporation is achieved.

Apart from neighbourhood relationships inscribed in the narrow circle of daily life, all the networks between people imply exchanges of messages by direct contacts after travelling, by indirect contacts entrusted to intermediaries, or by means of postal services or messages transmitted by electrical or electromagnetic waves. Goods and money are transported and information transferred. People come and go. The functioning of social networks depends on the technical quality of transport and communication networks which constitute the infrastructure. The performance of networks is higher when travel is easy and the transmission of messages instantaneous. But the distance separating partners is always a handicap.

Social relations and distance

The major problem in the experience of relationships is the confidence that one can place in partners. As long as one lives beside them there is scarcely any concern: one witnesses their actions and reactions and one understands their meaning. Their attitudes and intentions are known constantly. One is immediately aware when a change takes place in their behaviour. It is possible to ask the reasons and to have a frank explanation for clarification.

This is more difficult at a distance. When an expected response is late a doubt is insinuated: are the persons one is engaged with still on speaking terms? Can one still be sure of their intentions? One tries to reassure oneself: one telephones, one writes. A soothing reply arrives. The correspondents profess their good faith, allege tiredness, overwork, the irregular postal system or a telephone breakdown to explain their silence. Doubt is not dissipated; a meeting must take place if one has the time and the physical means. But it is costly. All the same, the efficacy of the contact network is reduced: by the cost of travel if one decides to bring things out into the open and by the hesitation to take matters further if the atmosphere remains strained.

When partners one deals with do not live locally, it is difficult to follow developments sufficiently closely to be certain of their intentions and good faith: it is the obstacle encountered by all social life when it takes place within a large area. To overcome distance it is useful to travel quickly and to have access to effective means of communication but this is not enough to create a good atmosphere: one must be sure of the partner's state of mind. Sometimes this is achieved by force, by abducting people dear to the interlocutors. States have practised this since antiquity, but this process is not possible for an individual. One is limited to demanding from partners solemn promises which one knows are only valid if there exists a minimum of trust. Adhesion to the same faith or to the same ideology therefore seem essential guarantees.

The institutionalization of relationship systems responds to these imperatives, but with uneven efficacy. Solidarity is strong within family lineages. Confidence is so natural between their members that separation does not alter relationships. The tie resists distance, but it remains limited to the family. Within a feudal system, the sworn faith guarantees willingness to act in common: the relationship is effective even when those united by it do not all live in the same place, but it cannot extend to infinity. The interests of those who meet at a market are spontaneously contradictory: each seeks to profit at the expense of the other. This reduces the circle where trade is possible and necessitates the meeting of all the partners and the goods they wish to trade at the same place: it is the only way to avoid being duped.

Regional analysis therefore cannot be limited to the examination of economic specializations: it must see by which networks contacts develop, businesses link up and solidarities develop. The brutal power of tyrants is ineffective in the sense that it can count on no one's loyalty: suspicion permanently undermines it – rightly so; the despot's agents only obey when they fear the brutality of his reactions: the further away they are the more they take liberties with the orders transmitted. The relationship of authority to power is infinitely more effective in modern political regimes. A government knows that its decisions are accepted by the overwhelming majority of the population and that it can count on the effective intervention of its administrations, the civil servants of which identify with the state. Since deviants, recalcitrants and opposers exist, a strict control over space is necessary. The analysis of the structuring of space in social relation networks must thus be complemented by that of surveillance systems.

Control relationships

Political life requires rather specific information flows; to take the pulse of the population, know the difficulties which assail it, to receive warning of the resultant discontent and frustrate the manoeuvres of enemies of the regime, governments need to control space and people. The situation is singular in the sense that those in the know and who hold information useful to the government do nothing to reveal it. To be informed, permanent observation and surveillance is necessary. Once information has been collected, its dispatch passes through hierarchical networks which, at each level, treat the entire data set, clarifying what might be learned and transmitting it upwards. But the difficult stage is data gathering.

Recalling the thinking of Jeremy Bentham two centuries ago, Michel Foucault underlined how control imperatives oppose others: when it is intended to put a population under surveillance, space must be compartmentalized into small units. The boundaries must be sufficiently clear and each cell must be sufficiently transparent so that those in charge of maintaining order can observe everything that is happening (figure 6.1). They thus immediately see any deviant behaviour. If control is constant, the certainty of being unable to escape the inquisitorial gaze of the authorities results in the disappearance of any whim of revolt or antisocial behaviour.

If it is wished to miss nothing that happens in a territory, it must be divided into clearly defined districts with agents charged with intelligence-gathering placed in residence in each centre. So that they miss nothing, they must have a total view of the whole area that they are responsible for. If this is not possible, informers are used who pass on immediately everything that they witness. The activity of Ministries of the Interior aim to achieve these conditions.

The compartmentalization resulting from the life of inter-relationships in general creates entities in which the poles and central areas are easily recognized but the margins of which generally remain blurred.

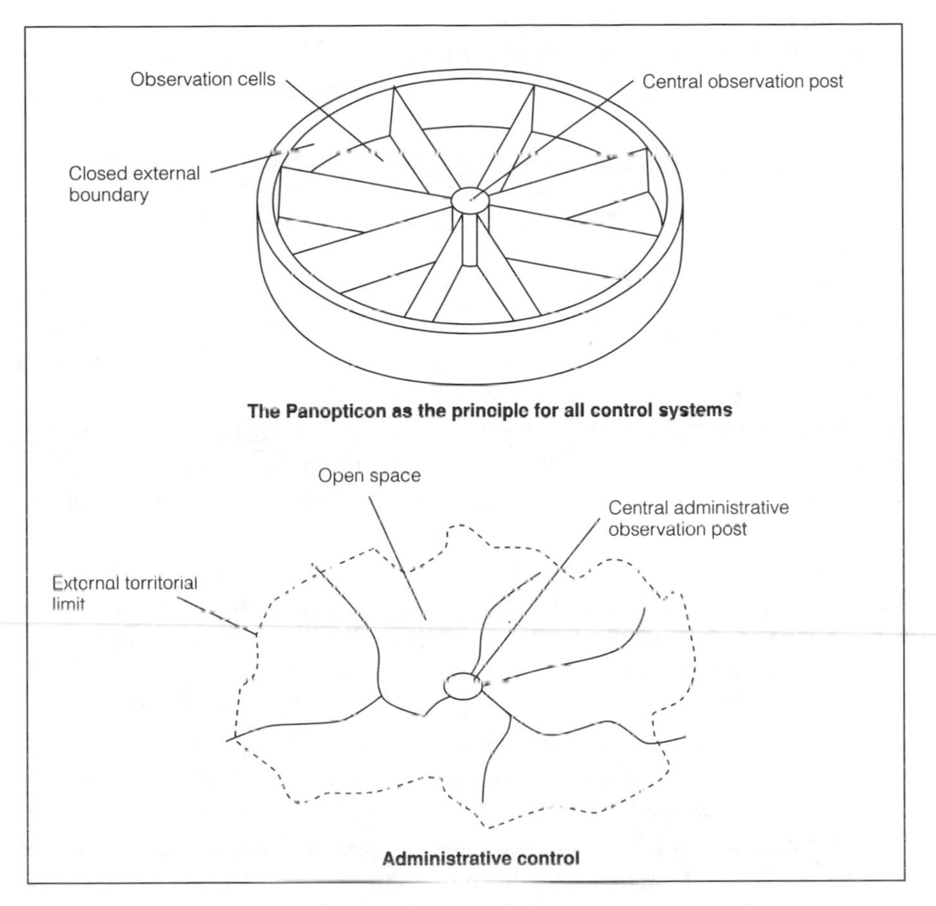

Figure 6.1 The logic of control and administrative structuring

Halfway between two centres of equal importance people are deliberately hesitant, visiting one or the other or both with the result that there is a zone of indecision rather than a frontier between their hinterlands. The compartmentalization required by a national administration cannot accommodate such indeterminate zones as they make control and surveillance impossible. To raise conscripts at the time of military service, everyone must have been included in a census at commune (or district) level and each commune must be attached permanently to a *canton*, to an *arrondissement*, to a *département* and to a region.*

The contradiction thus expressed between the needs of relationships in life generally and those of the administrative framework, linked to control and surveillance indispensable to all political systems, partly explains the ambiguities in regional policies. In the name of administrative rationalization, countries are given districts traced on the

* This classification is based on the French example of the local government hierarchy.

hinterlands of towns. This division is scarcely in place when it appears as an often intolerable constraint for those individuals who would prefer to remain masters of their choice as to which centres they frequent.

6.3 Transmission of Cultures, Segmentation and Differentiation of Space

The life of inter-relationships thus leads to an organization of space distinguished by specialization in zones of production, division into service areas and nested administrative districts. They are controlled by central places which often coincide but their boundaries (lines or zones where influences are shared on the one hand, linear axes on the other) do not always coincide nor are they of the same character. The segmentation of space results also from more fundamental aspects of group life: those linked to the transmission of cultures.

In animal societies, behaviour is essentially controlled by instinct. Learned behaviour counts less. In man, thought and action always takes place within frameworks transmitted by previous generations. Biological impulses exist but are not directly translated. The rules, knowledge and codes that guide them are assimilated in childhood and progressively modified and amended due to the ingenuity and creativity displayed by people when confronted with unexpected problems.

Culture is what we learn without realizing it, by imitating what happens around us and by what we have acquired during the periods of initiation, apprenticeship or schooling, which were imposed on us. It is also what personal experience, our imagination and our intellectual and moral faculties have permitted us to add – usually inspired by the example of our contemporaries. It is by this ensemble of knowledge, practices and norms that human societies exist and define themselves.

The manner in which knowledge is transmitted plays a capital role in the shape of cultures and in the way in which they are distributed over space. What results from the initiative of each individual weighs much less than the heritage.

Speech and spatial differentiation

The training of children commences from the youngest age: a baby learns about the world by catching hold of objects, carrying them in its mouth and later by moving around. It discovers that the environment is not uniform, penetrable here blocked there, and that it must beware of the angles of furniture. The acquisition of vocabulary is achieved by listening to those around it naming things. Learning the basic bodily discipline takes place at the same time.

A child loves to play. It watches adults and loves imitating them: sometimes it is the father, sometimes the mother, sometimes the doctor who takes its temperature and gives injections, sometimes the cook who prepares meals. It progresses unknowingly from games to work: to amuse it, one gives it a few peas to shell; one day it will be seated alongside its mother or father to help properly.

Before the invention of writing, the transmission of cultural elements was always based on imitation, advice lavished by members of the entourage and gestures to guide the acquisition of skills. The process is long. Childhood is devoted to it without there being special times which could be defined as being restricted to this process only. The formulae which must be learned and recited by heart show however that the process is never purely spontaneous. Its institutionalization and codification become apparent latter, at the time of adolescence. Before attaining adulthood, the young man or woman traverses a period of intensive instruction involving retreats from society, trials and initiation rites. The latter have the objective of ensuring, through their solemn, and sometimes dramatic magnitude, the interiorization of what seems most precious in the collective heritage: rules of behaviour, morality, the sense of duty and devotion to the common good. At the same time instruction in an occupation becomes more formal. To master techniques fully, it is not sufficient to observe what is done around one: one must be placed under the authority of a specialist who takes care to explain the basic manipulations, to analyse them and to have them repeated until they are reproduced instinctively.

In societies with oral traditions, the transmission of cultures is a purely local process: one must observe and listen to imitate and learn. Knowledge does not travel beyond those who are its repository, so that if one wishes to give one's son knowledge that no one in the neighbourhood has mastered, he must be put into apprenticeship elsewhere and change his teacher several times so that he can compare and progressively fill out his capital of knowledge. It was to meet this need that in the olden days, the *Tour de France des Compagnons** responded.

Cultural transmission takes place in a restricted space in societies without a written language. Knowledge, attitudes and values are learned locally or within a very small radius, by word of mouth, by gesture and by imitation. This introduces a tendency towards pulverization. The Kanaks of New Caledonia today speak twenty-eight languages (at the time of the first contacts with Europeans as many as thirty-two existed), even though they scarcely exceed 60,000 and have never been much more numerous. The installation of this population goes

* An allusion to the practice of itinerant study with master craftsmen, the *Compagnons*, who were members of specific trade guilds.

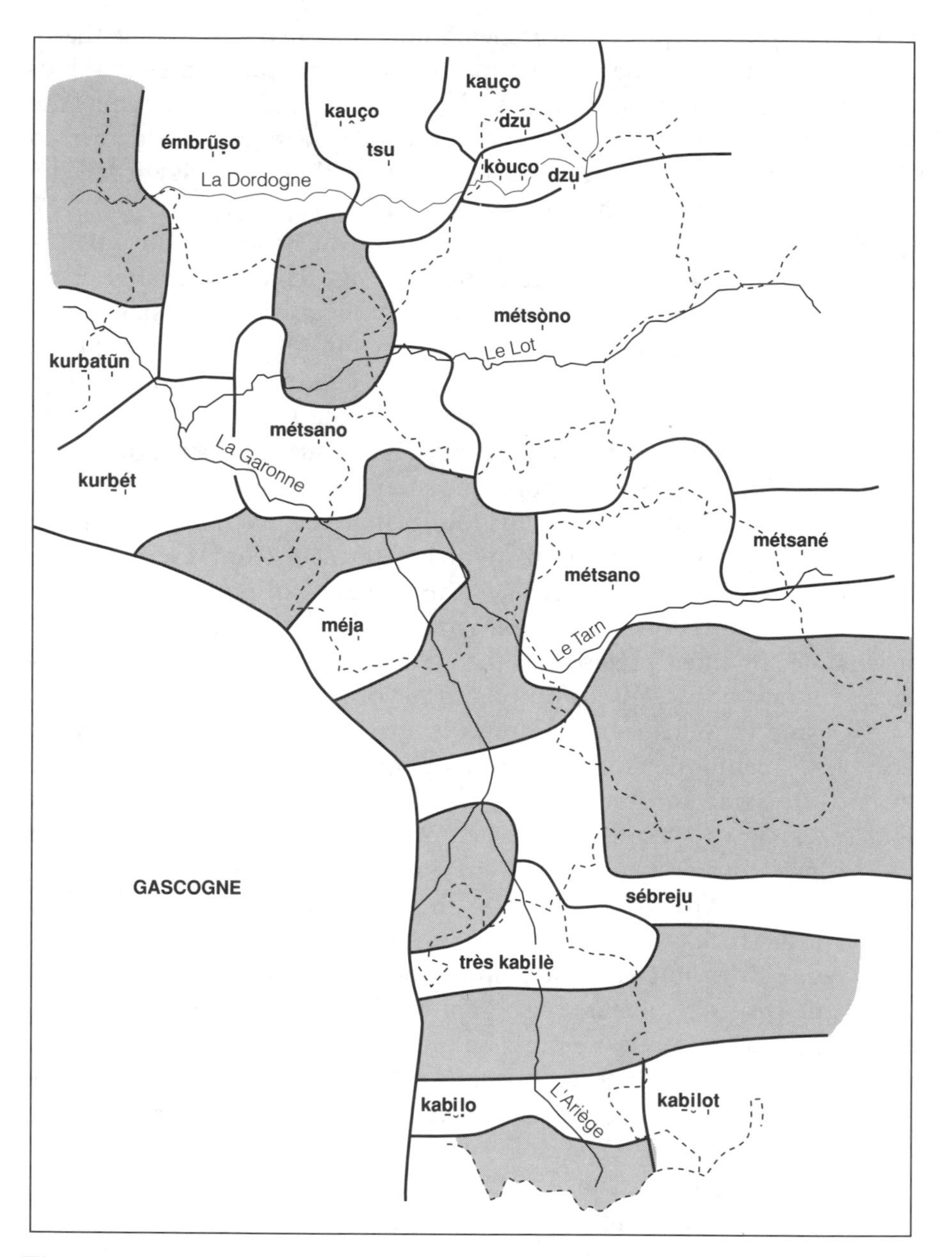

Figure 6.2 The marquetry of dialect variations: the designation of the word for the supports for animal yoke rings in the western Languedoc The terms and their pronunciation change over short distances without it always being possible to detect clear distributions (mixed zones are shaded). This tendency is especially frequent for technical terms.

Source: Atlas linguistique et ethnographique du Languedoc occidental, map II 520.

back to a very distant past. Nothing later has brought about a forced merging of tribes which were progressively constituted: local divergences have had plenty of time to become amplified.

It is in the realm of the techniques of material life that differentiation is especially asserted: the manipulations and tools used to produce everyday objects diversify from one place to another along the generations and give rise to a multitude of local traditions. Cultures vary in this sphere but in an uncertain manner; when their characteristics are mapped, a patchwork is obtained rather than a distribution of large units. The direction of knowledge and attitudes by word and gesture does not lead to the constitution of regions but rather to a crumbling which defies regrouping.

When mapped characteristics display continuous homogeneous ensembles, it is via factors other than a local transmission from generation to generation that an explanation must be looked for. Often it is the history of the population that is involved. People who spring from the same roots retain many common characteristics. Ten generations at least are necessary before intercommunication ceases to be possible. We measure this by seeing how languages from the same branch become differentiated. In Africa, the Bantu, bearers of an agricultural civilization and techniques that allowed them to prosper in the savannas and, later, in dense forest, conquered all the southern part of the continent from a heartland located in present day Cameroon. The process began in the first millennium before our era. The Afrikaner colonists were present at its last phases when they encountered along the river Kei, in the course of the years 1770 and 1780, the migrations of the Ngonis who had managed to occupy the relatively humid eastern part of South Africa. These people, whose dispersion thus began a long time previously and took place over immense areas where travel was difficult, speak languages that remain relatively close to each other.

The neolithization of Europe, in the view Colin Renfrew advanced recently, took place by analogous processes. Thanks to the superiority conferred on them by mastery of agriculture and animal rearing, the Indo-Europeans dispersed in all directions from a heartland that he believes lay in Anatolia. In this case the migrations must have begun in the fourth millennium before our era. Linguistic fragmentation is much more developed than in the Bantu area – this is normal for an evolution that was almost twice as long.

Not all the techniques employed by primitive societies are of local origin. Some arrived with the first settlers, others result from diffusion from an innovation hearth of domesticated species, tools or methods of working. Their advance depended on imitation and discourse. Travellers easily propagated the idea that certain projects are possible, but that is insufficient. One only decides when one has seen it, on the occasion of a meeting with neighbouring groups, or when one of their members moves in. Transmission is achieved by small steps rather than jumps. It gives birth to continuous and homogeneous zones, and thence to regions.

In a given place, the quest for effectiveness often means having recourse to the same procedures for hunting, fishing, pastoralism or agriculture, which in the long term influences the outline of diffusion zones. The areas where the same techniques are practised end up by coinciding with environments to which they are well suited. From one end of New Caledonia to the other, and the Melanesian domain, group life depends on the same tubers, taros and yams, and the same methods are universally used (the earth is banked up around the stalk for example, to facilitate the root growth). It is in pre-Columban America that the ways of life coincide best with the major natural divisions. In Africa, we know, after the synthesis expounded by Bauman, that cultural areas have a circular organization. Within the continent's major natural zones, a fragmentation has occurred around the hearths from which innovations diffused. Thus we can distinguish, in the Sudanese zone of East Africa, a circle of Upper Niger (from Senegal to the Bamako region), a Volta circle (from the Volta to the Niger) and the circle of Central Sudan (to Nigeria, Chad, Cameroon and Central Africa). The environment controls the diffusion of techniques, but in this case is not the only factor in play.

Popular cultures and elite cultures

From the moment when societies master writing, the transmission processes diversify. In infancy and, so far as the majority of attitudes, gestures are concerned, up until adult age, nothing is changed: it is still by looking around itself and copying what it sees that the child develops. But for morality, philosophy, religion, the rules of behaviour and knowledge in an abstract form, written text forms a satisfactory support. Culture from then on ceases to exist only in the memory and in the behaviour of its bearers. The knowledge owned by groups is transmitted in writing. This transforms it in several ways. The invention of writing did not do away with traditional modes of knowledge transfer, it complemented them. In the first stages, children and adolescents continue to depend totally for their education and instruction on what is said and done around them. Writing served simply to facilitate administration, to draft accounts, to carry out the computations imposed by the use of a non-lunar calendar and to write out sacred texts, but did not in any way modify anything related to the technical realm.

From now on culture no longer evolved on the basis of collective memory. The texts where certain of its aspects were transcribed rendered them timeless and permitted the authority of the patriarchs to influence their present-day descendants. Having become an easily transported physical object and, after the invention of printing, cheap and easy to reproduce, culture travelled easily. It was no longer necessary to travel to learn what was being said or done elsewhere. The abstract elements and contour of culture thus became easily transmissible beyond the immediate environment.

For children, recourse to the written word is not directly possible: they must first be taught to read and write. Access to books containing new knowledge or the rules of behaviour established by the revered authorities is not easy for young mentalities. Schoolteachers and professors teach the content of books. What has been concretized in writing and thus permits diffusion, they transmit in part orally and by learning, as in the most traditional modes of education, to recite canonical texts learned by heart.

The duality in the modes of transmission very quickly translate into a marked divergence between knowledge accessible to the majority, the education of which consists only of words and actions, and those with the knowledge of the elites using books. From now on, culture has two levels: its mass component is responsible for the majority of production techniques and often with a very rich fund of mythical recitations and religious beliefs, attitudes and moral principles; its elite side is also turned towards religion and morality, but is presented in a more systematic way, based on revelation or on great metaphysical or philosophical texts; it lends itself more to techniques of social control than production; it assumes a prominent place in artistic creation.

Intermediate societies owe their distinctive spatial characteristics to the dual cultural dimension. The fragmentation of mass knowledge remains in evidence. In the linguistic field this is translated into a proliferation of dialects; vocabulary, grammatical rules and expressions often change over very short distances, even though the areas related to each characteristic overlap to the point that it is difficult to attribute linear boundaries to this or that variant in language. In the technical sphere, the contribution of local processes of transmission are visible in the variety of tools and techniques, the architectural details, the way houses are decorated or the form of furniture and men and women's costumes.

Elite culture is available within a framework that extends over much larger areas; writing permits the same language to be used at widely separated points; it facilitates the rapid diffusion of beliefs, religions or ideologies, and quickly ensures the sharing of the results of rational thought and scientific research by large communities. A double structuring of space corresponds to the two cultural levels: narrow local cells for the mass level and much larger extent for the informed culture. The latter's dimensions are highly variable: sometimes they stretch over a whole continent, or they may be restricted to a relatively narrow linguistic zone or to a nation conscious of its common origins and the distinctiveness of its destiny. When movements are too uncommon to maintain direct mutual comprehension of the spoken language, the unity of the educated culture is safeguarded by the use of a dead language: Christian Europe owes to Latin in the West, and to Greek then Slavic in the East, its escape from the fragmentation that the breakdown of the Roman Empire and the barbarian invasions encouraged. The opposition between the Greek and Roman factions was maintained and hardened when the schism definitively destroyed the religious unity of the continent.

Recourse to a dead language accentuates the elite character of a culture, but inevitably the moment arrives when a return must be made to a living language for contacts within the ruling classes. The educated culture zones then fragment and assume hierarchies: in medieval Europe, the unity was linked to Latin, but national literatures in the vulgar language appeared one after another and use of the speech of the court finally was imposed in all proceedings of public life. It happens that a foreign language may be turned to, to enlarge the area where educated culture is received: French played this role in part of medieval Europe, and from the eighteenth to the twentieth century.

The two cultural levels are not impermeable. It is true that the relations between the ruling culture and popular culture may be minimal. The elite often speaks a language incomprehensible to the majority of the peasant classes. The latter belong to different ethnic groups. In the former Bukovina, the large landowners possessed villages populated by Ruthenians, Russians, Ukrainians or Romanians. In Bohemia, the jigsaw puzzle was composed of Germans and Czechs and was dominated by the Germans; in southern Slovakia, Slovaks and Hungarians from which the ruling nobility were drawn were mixed. Transylvania juxtaposed communities of Romanians, Hungarians and Saxons (German-speaking). But even in these cases, the upper levels facilitated and shaped the diffusion of innovations. Whilst the most diverse of languages coexisted, generally the same religion and faith was practised. The unifying action of higher culture was most evident at the level of ideologies and moral and political attitudes.

Techniques were absent from the ruling culture of the elites for a long time. They did however play a role in their diffusion: within the units guided by the same morals and codes of behaviour, it was relatively easy to travel. Masons and stonecarvers moved from one building site to another; cabinetmakers settled wherever they were given orders for furniture. The organization of apprenticeship itineraries, the *Tour de France des Compagnons* for example, were inscribed in the same logic.

The role of educated culture has other facets. Highly varied objects became associated with particular countries thanks to it: architecture, furniture or costume assumed a symbolic value and were adopted, within very large territorial groupings, by the elite cultures. This does not shape the technical scene, but intellectuals exploited it to cement solidarity and enhance the image of territory-building.

Spatial structuring of intermediate societies thus takes place at two levels: a narrow segmentation for what is transmitted by imitation and oral contact, and larger units for more abstract realities. In the first category belong the majority of technical skills (they are poorly transmitted in writing while their formulation in a scientific manner has not been learned). To the second falls science, philosophy, religion and the major ideologies on which societies base their coherence.

With printing and the generalization of schooling, the boundary between mass knowledge and educated culture was moved. The knowledge of the humble classes draws ever more widely on the intellectual resources that scientific progress opened up to include techniques; popular culture ceased to be regarded with suspicion. As evidence the place given to fairy tales, popular songs and tradition in literary, musical or artistic creations from the eighteenth century onwards may be cited. A fundamental difference remained however: popular culture continued to be dependent on oral transmission procedures. Its bases were still limited within restricted areas.

The impact of means of mass communication: mass culture and specialized technical cultures

The ways of recording, conserving and diffusing information have multiplied. Today we can record the word, take photographs and make films. To the use of paper, universal until now, others have been added: film, electrical or magnetic impulses, engraved disks etc. The speed of transmission has accelerated and in many cases is an instantaneous process. At the same time, scientific progress permits the tackling of technical problems with new tools: instead of involving manual skills, practical understanding springs from the application of scientific laws. The difficulties encountered in long-distance transmission have disappeared: one can learn by reading and, with the aid of specially made films or tapes, acquire any necessary manual skills.

These changes, which began in the last century, have accelerated prodigiously in the last fifty years: television brings images from everywhere into the home; a variety of recording systems allow voice and image to travel easily; telecommunications systems guarantee instantaneous transmission of extremely dense messages. The role of parents and their immediate entourage has not disappeared from childhood, but a major interlocutor appears early: the television set. The voice and actions that the young copy are no longer only those of the immediate environment: a large part of what used to happen locally from generation to generation finds itself replaced by messages from distant places. We leave behind the age of popular culture for that of mass cultures.

We are still not fully aware of the impact of new means of communication on the structure of contemporary activities. What has abruptly declined are the old locally transmitted resources of common knowledge. With them disappears part of the technical skill, the methods of cultivating and managing rural farmholdings, to run a small artisan business or to maintain a household. Cultural acquisition today consists of watching television, listening to the radio or tape recorder. What is thus learned is to amuse oneself and to participate in shows organized by others. By watching the life of stars on the screen, one is overtaken by the desire to behave like them: one observes what they consume and the way in which they entertain themselves and tries to imitate them.

The new means of mass communication could demonstrate how to learn new techniques because the skills are demonstrated visually. But our instrumental world is so diverse that the pathways of learning are multiplied: to ensure a large clientele, a public television network or a company producing records, cassettes or diskettes has a vested interest in an orientation towards whatever reaches the largest possible number of people. In this sense, mass cultures become cultures of consumption. The old tendency towards fragmentation into small areas of the apparatus that we surround ourselves with is effaced. In many domains, the natural inclination until the last fifty years was diversification. Today the trend is towards uniformity.

The part of culture the transmission of which depended more on writing and schooling (the informed culture one might say) had wide territorial foundations and was more uniform. By virtue of the religious or ideological content, it tended to consolidate all those receiving it in the same form. The transmission facilities of educated cultures progressed in parallel with those benefiting popular culture. Today, to be introduced to a foreign language, manuals (and the exercises they contain), tapes (which familiarize the listener with pronunciation, the position of the tonic accent and the rhythm of sentences), films (which associate the word with the movement), and the reading of newspapers and books, are combined. It is much easier to learn without a teacher because a series of aids to help the comprehension of the diverse aspects of complex messages is available. No longer is it a privilege associated with high social standing to have access to educated cultures. These suddenly lose part of their function in the formation of hierarchies. Their content moreover changes: they turn towards more technical matters. They are designed to teach those who so desire an introduction to a capability; they are no longer oriented towards the transmission (of necessity, abstract) of religions, philosophies or ideologies. This becomes only a subsidiary role.

While the advent of mass culture tends to homogenize societies and the space shaped by them, the transition of the elite cultures of the past to the specialized technical knowledge of today reintroduces a certain diversity. This, however, does not follow the same pattern as in the past. Modern means of communication permit people living distant from each other to remain in continuous contact. Those with similar activities, tastes, recreation or aspirations no longer need to live next to each other. They do not form continuous territorial wholes. However, at the local scale, profiting from the possibility of daily travel, it transpires that those who share the same interests and enthusiasms and live in the same agglomeration, locate in the same district, or in the evening frequent the same streets, restaurants, or even the same meeting-rooms.

Modern means of communication facilitate the constitution of networks of specialists, but the technical cultures which replace the former educated culture do not contribute as much to giving society a unified structure. They often tend towards segmentation, but a segmentation which does not have a geographical base and does not impede global society from continuing as a place of exchange. What is new is that it is decreasingly founded on a shared enterprise.

Conclusion

Cultural factors are therefore central to a division of humanity into distinct societies. In primitive humanity, where transmission is supported only by words and gestures, the tendency is for division into small, practically autonomous units. In literate societies, transmission takes place by a double process which maintains the explosion of the technical universe within small cells and favours the emergence of much bigger units which communicate via the same values and endeavour to unify their languages. Modern means of communication lead, through mass cultures, to the uniformitization of consumption and related behaviour. Specialized knowledge multiplies, but rather than giving to the groups that share it reasons to live in common, it tends to isolate them from each other even when they cohabit the same space.

The segmentation linked to the transmission of culture, changes its meaning on passing from primitive humanity to literate societies. The compartments that share the same oral cultural cease to appear as being watertight from each other: they are overlain by vaster spaces unified by religions, ideologies, philosophies and shared rational knowledge.

With mass culture, forces encouraging unification exert themselves and the traditional compartments are effaced. Cultural transmission is only one of the essential factors in the segmentation of space, even though increased communication facilities favour, by widening everyone's choice, the multiplication of the individual's options.

Faced with a world that has become too similar, reactions are becoming apparent, however: to affirm their identity better, groups attach themselves to ideologies that make the most of practices, attitudes, knowledge or beliefs that differentiate them. The fundamentalist vogue is merely the extreme manifestation of this.

The modes of cultural transmission weigh directly on the structuring of societies and on the distribution of those receiving the same heritage. However, it is exceptional that the resultant areas have a regional dimension. In certain cases it is the tendency towards pulverization that gains, in others it is the uniformitization. The ensembles that share the same heritage are not always continuous – this is particularly true today. The diversity of territorial units, which strongly encourages the birth of collective identity, owes much however, to the influence of local transmission techniques, know-how and attitudes. This remains true in the age of mass culture: in this case it is no longer popular culture which is in question, but the extraordinary blooming of small groups, chapels, circles, brotherhoods or associations made possible by increased mobility, better general education and modern facilities of long-distance communications.

Further Reading

Barth, F. (ed.), *Ethnic Groups and Boundaries: The Social Organization of Cultural Difference*, London: Allen and Unwin, 1969, 153 pp.

Castells, M., *The Rise of the Network Society*, Oxford: Blackwell, 1996, 480 pp.

——, *The Informational City, Economic Restructuring and Urban Development*, Oxford: Blackwell, 1991, 448 pp.

Foucault, M., *Discipline and Punish: The Birth of the Prison*, London: Allen Lane, 1977, 333 pp.

Graham, S. and Marvin, S., *Telecommunications and the City*, London: Routledge, 1995, 456 pp.

Goody, J., *The Interface between the Written and the Oral*, Cambridge: Cambridge University Press, 1993, 323 pp.

Gregory, D., *Geographical Imaginations*, Oxford: Blackwell, 1994, 442 pp.

—— and Urry, D. (eds), *Social Relations and Spatial Structures*, London: Macmillan, 1985.

Harvey, D., *Justice, Nature and the Geography of Difference*, Oxford: Blackwell, 1996, 496 pp.

Redfield, R., *Peasant Society and Culture*, Chicago: University of Chicago Press, 1955.

Seaman, D. and Mugerauer, R. (eds), *Dwelling, Place and Environment: Towards a Phenomenology of Person and World*, Dordrecht: Martinus Nijhoff, 1985.

Sibley, D., *Geographies of Exclusion: Society and Difference in the West*, London: Routledge, 1995, 224 pp.

Tuan, Y.-F., Space and place: a humanistic perspective, *Progress in Geography*, vol. 6, 1974, pp. 211–52.

——, *Topophilia: A Study of Environmental Perception, Attitudes and Values*, Englewood Cliffs, NJ: Prentice-Hall, 1974, 260 pp.

7

Regional Consciousness and Identity

People are not robots. Their decisions reflect past experience, the values transmitted to them and those adopted later. The environment, the conditions of its exploitation and the technical level in general are not the only factors involved in the earth's regional differentiation. What changes from one cell to another is the way in which the world is perceived, the understanding of nature, the role accorded to it and the significance accorded to the individual and to collective destiny.

Recent research has been concerned to demonstrate the way in which the transmission of knowledge helps or inhibits regional diversification. It probes everything that the structuring of experience and exploitation of space owes to the significance that people give to their existence, to the natural environment and to the world in general.

7.1 Life-world* and the *Pays*

The life experience of space

The introduction to life is simultaneously the discovery of the world. By learning to walk the baby learns that the setting in which it is growing up is full of traps. The surface it walks on is almost perfectly flat but is interrupted by a flight of stairs down which he will tumble if he negotiates them carelessly. By trial and error, the child learns the diversity of objects filling up space, their colour, texture and the way in which they can be pushed, overturned or moved.

The experience of the garden and the local district are little by little added to that of the house. Through movements with adults or friends,

* See translator's note on p. 22.

the child becomes used to finding out where it is unaided, finding its way to a neighbour's or the grocer's. Perception becomes organized and takes place differently according to the environment. In the countryside one is oriented by a tall tree, the village steeple, such-and-such a farm, the tracks that snake across the fields or footpaths and permit one to wander through the woods without getting lost. In a town with traditional architecture, the image is made up of the corridor bordered by buildings constituted by streets; the problem is to remember the way in which they intersect. Where the buildings are isolated and disposed without respect for alignment, perception is organized a little as though in the country: so as not to get lost, one must memorize the paths crossing the lawns between housing blocks and using the towers as signposts.

The image that people construct of their environment in this way reflects the forms that space has received. It develops through the rhythmic movements of life: it has a functional character because it incorporates the journeys leading to the school, shops, church, sports ground and, later, to the workplace. These to and fro movements are equally motivated by family ties and recreation.

In becoming familiar with places one meets people. The face, name and background of people one meets often are well known. Elsewhere, it is a vaguer impression: one identifies certain faces and feels at home, even in the middle of a crowd, since these are always the same kind of people, with the same attitudes and behaviour of one's setting. One knows in advance how people will react since they share the same values and demonstrate the same feelings: what makes you laugh makes them laugh, and what seems serious to them is the same for you.

The world that one frequents daily is so familiar that it is inbuilt; memories are everywhere associated with it. The culture transmitted to us is not made up of abstract and disembodied messages. It has been progressively revealed by people who make it concrete, close and living. It is in the farmyard, the village centre or on the pavement of a suburban district where one learned to walk, run and play. Women are initiated into cooking, knitting and sewing in the room where their mother bustles between the stove, the ironing board and the sewing machine. Men owe to their father how to use tools in the garden, fields or workshop. Thanks to the local mechanic one knows how to take apart and repair a motor. Each experience has its (or their) site(s) and evokes specific memories. The child knows where to buy the sweets he likes, the young woman knows the fashion boutiques, the hairdressers' or beauty salons, the men know the DIY stores. A whole series of connotations are linked to effort and work: it was there that I went to school, here is the office or factory where I worked, the shops where I bought furniture, the clients I visited, the bank that gave me credit. But the strongest colorations come from the places where the decisive stages in existence took place – the playground where pals were made, the dance where one met one's wife, the church one was married in, the cemetery where one's parents rest.

The familiar universe becomes enlarged with age and diversifies with one's employment. In passing from kindergarten to the primary school, to junior and senior school, and then to university, one climbs a whole hierarchy of ever more distant locations. In adulthood, those who work, almost all men and an increasing number of women, structure themselves around the two poles of home and work, and the house remains the focus of housewives, children or retired persons.

After a certain age, mobility decreases. People live within a more restricted sphere. Their minds retain the wider horizon that they scarcely visit any more. They do not see its evolution directly and are only aware of its change by chance conversations, or mention made in the press, radio or television of building sites opening or facilities and monuments being inaugurated.*

The experience of home and the feeling of pays

A feeling of familiarity with places and people thus develops, expressed by saying that one feels at home or, again when one speaks of one's *pays*. All languages have an expression for it in one way or another; the family space is evoked by 'home' in English-speaking countries, and by 'Heimat' where German is spoken. The term refers to the whole of what one knows well and where one is known and recognized by all. The extent of this fundamental compartment of the life-world varies with people's mobility. It is a function of place size. In major agglomerations the familiar universe includes the cell where one lives and relatively distant districts of the workplace, where one shops or goes for entertainment. The known area swells and forms a hierarchy: at the base is the restricted area of the local district, the only one known to children and frequented by old or handicapped persons; higher up is situated the often discontinuous ensemble of places frequented in earning one's living. Whilst the local cell is different for everyone, the zones where employment and businesses are concentrated are shared by many. Thus it is around the town centre, or sometimes a suburban commercial centre, that the image is built of places with which one is totally familiar.

* The process proposed above is to some extent a stereotype of past generations rather than a reflection of contemporary conditions but it here serves as an example of the cumulative nature of the construction of a life-world.

The first articulation of space that the geographer encounters thus springs from people's daily experience. It combines connotations familiar to all but which are never identical. Places evoke feelings but not in the same way in terms of age, sex and occupation; they depend also on the journeys that take place regularly over the year. Some people simply have a fixed base because they have always lived in the same place or because they have broken all ties with the environments they passed time in in the past. Others have several bases.

Those who live, or have lived, within the same frameworks share an experience, expressed by the saying that they belong to a *pays*, that they are from the same *pays*. This is the meaning that we must attach to the regionalist slogan 'We wish to live and work in our *pays*'. Research by historians, sociologists and ecologists in the last fifteen years speak in the same way, of *pays* as the shared space experienced in everyday life and in daily or weekly journeys.

Multiple attachments

Populations have never been totally sedentary. In olden days distant fairs were visited. The great pilgrimages mobilized crowds; the horizon was rarely as circumscribed as might be supposed. But for the majority of people, major journeys were too exceptional for the familiar space to extend beyond the nearest markets. The situation was a little different for women obliged to change their residence at the time of marriage. they had two areas of attachment but did not live in them in the same way: there was that of their memories of childhood, parents, brothers and sisters, and that of the husband, work, children and in-laws. One recalls with emotion the places of one's youth, but thereafter one belonged to a new universe.

The habit of country retreats is an old one among the wealthy classes. Goldoni dates it from the middle of the eighteenth century. Noble families spent the summer on their lands and the inclement season in the town. For members of parliament or the wealthy bourgeoisie, the main residence was urban, but the country lands acquired in the quest for prestige, where one retreated during heatwaves and where wife and children spent a longer period, also constituted a place rich in memories and multiple emotions. The wealthy classes thus were rooted in two locations. With the general application of paid holidays, this phenomenon has extended to involve a very large proportion, sometimes a majority of the population in developed nations.

Areas where one feels at home and where one is part of the *pays* belong to the common experience of humanity, but to map them is impossible: there are as many centres as there are people, and their boundaries change with everyone's age and generation. It is frustrating for geographers: they feel how important the reference is to familiar places to understand the relationship between individual and space, but the subject has eluded them as long as they wished to consider only territorial features with clear limits, a stable content and an objective status. Since 1960, the situation has changed as the interest in perception, the experience of life space and mental images of the world has developed.

7.2 The Collective Dimensions of Regional Feeling

Multiple individual experience of places coalesce from many points of view. Society has at its disposal thousands of means to standardize and normalize them.

Place names and the baptism of the land

The structuring that each person constructs of space is facilitated by the baptism of essential landmarks. Space cannot be known, shared and memorized except by language. The first concern of an explorer discovering a remote terra incognita is to name it and to give a designation to all its characteristic and pinpointed places.

Everyone learns to know their environment by reference to place names, well known to all their users. The names serve first to identify places: a farmer's wife tells you that her husband is working in 'long stripe', a field at the bend in the lane. The Michel family lives in Bérarde and the school is at Talmont. In town, one would know that the dairy is in the rue du Départ and the bakers in the rue du Lycée or rue Wilson.

The music of place names is associated with the experience of *pays* and brings memories flooding back. To the extent that families of place names exist, the evocation of a series of localities situates the environs where people live. Novelists know this: when in a few lines you learn that the hero has left his home in Calinac for the farm of Triquedina, and stopped at La Devèze on the way, you are transported into the depths of the Midi, in the Gers or Lot-et-Garonne. The town, which is at the heart of many novels, changes according to the scenes that are unveiled between the market place, the town square, the rue des Consuls or in the maze made up of the streets named after Gagarine, Barbusse, Vaillant-Couturier or Stalingrad in some industrial region.

The names do more than simply identify places. They bear varied connotations, which are social and geographical when they reflect the natural, linguistic, cultural and religious environment in which they are used, and political when they proclaim the ideological sympathies of the town council in power.

Many familiar place names are archaeological documents and conserve the history of ancient peoples who still haunt places. Very few French people know the Gallo-Roman origins of place names ending in -ac, -argues or in -y. Normally, no one pays any attention. What counts, however, is the resonance of words associated with places. Names that are considered to be French in reality go back to pre-Gaul or Gaul, to the Gallo-Roman epoch, to the great invasions, to the Middle Ages or to modern times. Whether they speak English, French, Spanish or Portuguese, the inhabitants of the New World are not inconvenienced by living in a country where the rivers, summits, sometimes villages are inherited from an Amerindian heritage. Australians and New Zealanders readily use terms recalling the first occupiers, Aborigines and Maoris respectively.

The geographical categories of experience in common

To the study of place names may be added that of more general terms, bearers of popular geographical knowledge. Triguedina is a farm, La Devèze a hamlet. The village is Calviac. To travel around in the countryside one follows footpaths, lanes or roads. In town, the highways are called streets, avenues, boulevards; they cross squares and interlink by crossroads. These terms aid the specification of space. Avenues and courtyards are generously planted with trees and are wider than streets. So are boulevards but they define circular enclosures that reflect the fortifications, the boulevards that surrounded the town in the past. Some centres merit the term city while others are only towns: the former were, and often still are, bishoprics; the Diocletian reforms located them in the Gallo-Roman capitals. The distinction is doubtless no longer necessary to understand the function of present-day centres, but it was significant at the time when vocabulary was defined; religious ties counted for much and the bishopric held a central zone of influence, command and control lacking in other population centres.

The notion of town centre elucidates the structure of urbanized zones, especially the largest but it is not always well conceptualized. Everyone knows where it is located in their home town and how to get there. When travelling, one quickly learns by analogy how to find the core of urban centres one passes through. It is identified by the presence of certain services, café terraces, high-order services, banks or administrative offices; within a specific society, this is translated in the landscape by distinctive architectural forms and spatial organization – *plaza mayor* in the Spanish world, the spiky silhouette of skyscrapers in the American context, *souks* in Arab countries.

The vision that we have of our *pays* is therefore not purely personal, even if it is loaded with memories and emotion. It is partly structured by what the use and application of words have taught us concerning the methods of spatial organization. It is remarkable how much the memories of features that have long since disappeared are perpetuated in vocabulary.

Among the words that we thus learn to employ, some have a double status: they have a general meaning applied to such or such a setting or they may refer to a specific place. In the Arab world, the *sahel* means the shore – and by extension the sometimes indistinct fringes between two environments, between the desert and the savanna to the south of the Sahara for example, or the Sahel of Algiers, constituted by the ring of hills to the west of the city, between the Mitidja plain and the sea from Tipasa to the Djebel Chenoua.

We are familiar, in the Jura mountains, with the difference between a *joux* (a pine forest) and a *pessière* (a spruce forest). These are general conceptions but often lend themselves as place names. Elsewhere, the term *ségala* is specific to acidic soils related to the outcrop of primary rocks: as they lack base elements they are poor and scarcely lend themselves to more than the cultivation of rye. Soils developed on *fromental* are richer, generally neutral, and crops of wheat are easily produced. Locally, as opposed to the *ségala* may be a *causse*: 'a shallow soil due to the breakdown of limestone', according to Fenelon. He continues: 'it is also an ensemble of soils and the underlying rock yielding land favourable to cereals and to certain vegetable crops'. Elsewhere in south-western France, the distinction is drawn between the *boulbènes*, 'silicious soils, composed of very fine sand, loëss and clay', and the *terreforts*, heavy, difficult to work but generally more fertile. In the Poitou region, the *groies*, reddish clayey limestone with rubble mixed in the soil, appear as lands well suited to wheat, the vine or vegetables. The *brande* lands, for long left to heather and increasingly to broom, gorse and ferns, correspond to clay-silica soils very poor in lime: they yield only very mediocre crops.

We could multiply to infinity examples of terms that define both a type of soil and the crops normally associated with it. By using them, we learn to apply to the environment a general classification matrix. Once this has been assimilated it can be transposed elsewhere.

The standardization of the experience of space and the names of pays *and regions*

From the names of types of land, one passes imperceptibly to place names: what is a *causse*, in the south of the Massif Central or in the

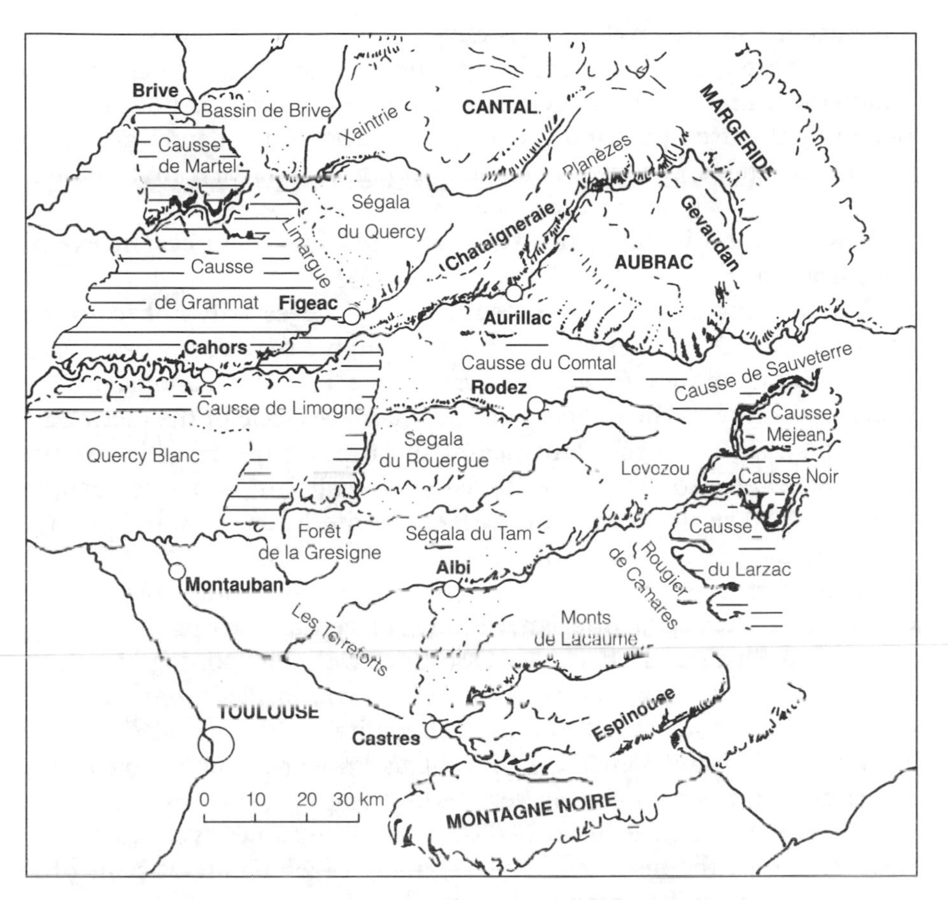

Figure 7.1 *Pays* names in the interior South-West of France

Quercy, other than a large plateau where the soil and subsoil correspond to the *causse* type just described above? Similarly, we speak of *ségala* to designate *pays* with silicious soils, whether in the Rouergue or Quercy (figure 7.1). In neighbouring Cantal, it is the dense cover of sweet chestnuts on the primary rocks that gives its name to the *chataigneraie*. In England, the 'downs' refers both to the chalk dry valleys with very short grass as well as to a regional unit – the belt of uplands dominating the vale of the Weald.

On Europe's Atlantic façade, it is the vegetation rather than the soil type that people have distinguished: the *lande* defines a terrain traversed only by wild species, heather, gorse or broom. By extension it defines certain *pays*, the *landes* in France, the 'moors' in England – Dartmoor or Exmoor in Devon and the Yorkshire Moors in Yorkshire.

When the name of a type of terrain or vegetation cover is thus applied to a small region or *pays*, the term conveys to the person encountering it two pieces of information: it evokes the idea of a certain type of soil or natural landscape and indicates the existence of a territorial unit. Even if one has never traversed it, if one only knows it from one place or short itinerary, one is aware that its agricultural endowment makes it homogeneous. Direct experience is not only enriched by the use of vocabulary, it is extended. Since names are given to regional tracts, the experience is multiplied. The *pays* that one has heard of is not known in the same way as by someone who knows all its byways, but it is sufficient to give an idea of its resources, its extent and location.

The meaning that names transmit are not necessarily linked to the environment. Often they refer to the inhabitants themselves or to their origins. This applies to many *pays* names: le Val d'Amour, in the Franche-Comté, is the old *pays* of the Chamaves, one of the Germanic tribes that settled in this province at the time of the great invasions. The *pays* Varais, to the east of Bescançon recalls an Alemanic* people who chose to settle there. The Lusace, in the heart of Brandenburg, indicates the persistence until today of a Slavic core.

Many of the *pays* names refer to units fashioned during feudal times: we can cite Quercy la Bouriane, a *pays* of verdant scarps and basins with small farmsteads between Gourdon and the Dordogne valley. Italy swarms with units of this type, Montferrat in Piedmont being an example. Others are structured by their capital – the *pays* of Vincenco, Treviso, Padua and Venetia – without us knowing how far back this usage goes, but which can be very old.

Beyond the limited areas formed by *pays*, place names evoke medium-sized divisions, the historical regions, many of which are ancient provinces: Burgundy, Languedoc, Savoy, Dauphiné, Normandy – names of former administrative divisions. In countries like Germany, these entities recall the composition of the Germanic peopling, settled in one part of the country or other: Freisans, Saxons, Franks, Alamans, Bavarians. The French provinces sometimes have retained the imprint of groups that settled or dominated them – the case may be cited of the Burgundians who created Burgundy – but usually in France one must go back to a more distant past to see what role the ethnic factor has played in spatial differentiation. The tribal territories of the Gauls, defined between the third and first centuries BC, became Gallo-Roman cities, then dioceses: the Auvergne, *pays* of the Arvernes, Berry of the Bituriges, Poitou of the Pictons for exmaple.

The designations employed to name these medium-sized realities are not all constructed on these models. Some recall the dominant influence of a major urban centre: thus we speak of the Paris region, the Lyon region, the Bordeaux region (which largely coincides with Aquitaine) or the Toulouse region. Others derive from academic divisions

* Refers to German-speaking Swiss peoples.

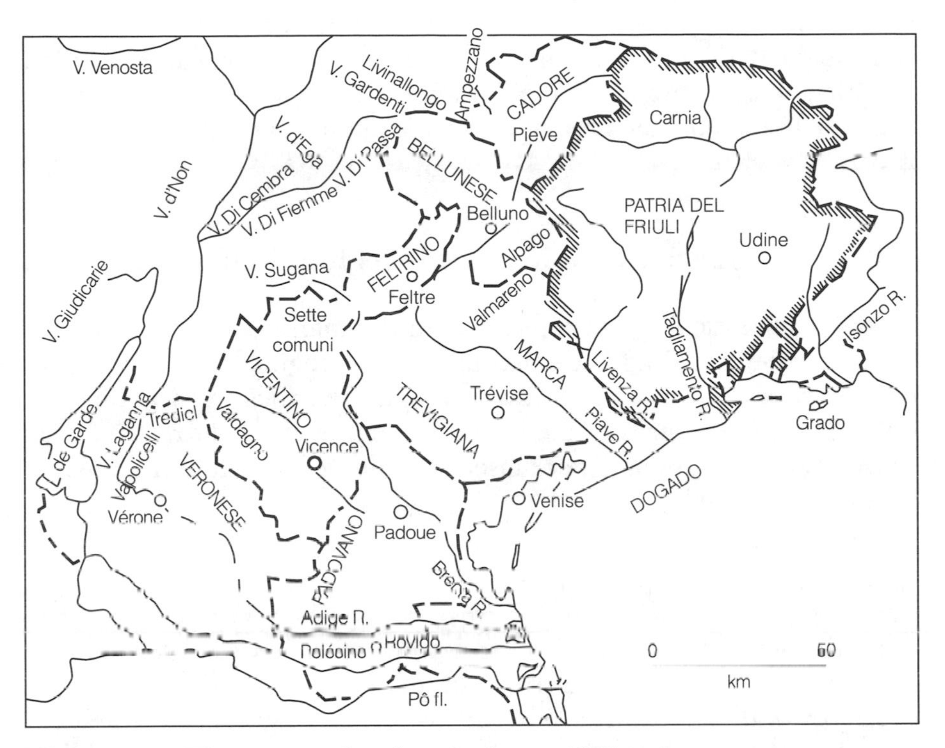

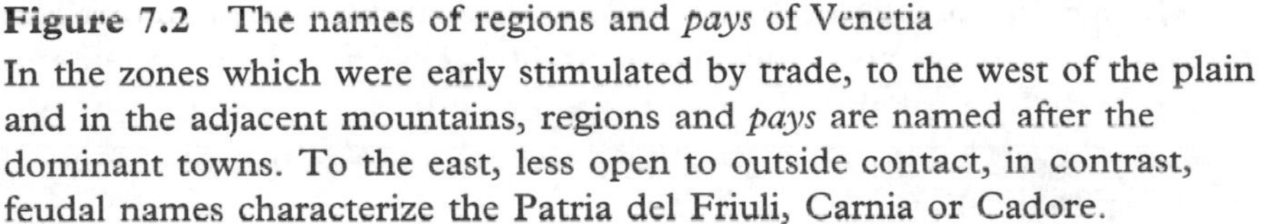

Figure 7.2 The names of regions and *pays* of Venetia
In the zones which were early stimulated by trade, to the west of the plain and in the adjacent mountains, regions and *pays* are named after the dominant towns. To the east, less open to outside contact, in contrast, feudal names characterize the Patria del Friuli, Carnia or Cadore.

which, since the last century, have become incorporated into popular culture – the Massif Central, a term invented by geologists, is often cited. In Canada, the term 'shield' is readily used, a geomorphological term for the rocky and forested vastness to the north of the productive zone.

Some regional designations are more vague: in Italy, the Mezzogiorno evokes the poverty, underdevelopment and the influence of the Sicilian Mafia or the Neapolitan Camorra: these are characteristics common to the islands and more than half of the peninsula. In France, one belongs to the Midi, the Nord or the West. These words do not have the same content: if one speaks of the Midi rather than the South, it recalls that there is an ensemble which differs from the rest of the country in accent, manners and its distinctive cultural and linguistic heritage. The Nord also evokes a strong identity, in spite of the linguistic split introduced by the use of Flemish. The East or West of France are vaguer entities, conglomerates of provinces rather than units felt as such. The use of these expressions became widespread during the railway era and recalls both the position of these spaces in relation to Paris and the name of the networks serving them. Pierre Monberg has followed up the formation of regional units similarly fashioned by the major railway companies in the state of São Paulo in Brazil.

7.3 Spatial Ontology and Regional Diversity

A social dimension always intervenes in the manner in which the real world is interpreted and regional divisions recognized. It is born out of what we have learned and draws on collective memory and on symbols. Alongside records of the banal, the flat or the unattractive, exist those of the exceptional, grandiose, sublime or authentic: the profane is contrasted with the sacred.

Ontology and the consecration of space

We have received an idea of the structures of the cosmos through our culture. It enables us to distinguish between the essential and the accessory. Deeper realities outcrop underneath surface appearances and change the nature of certain elements, beings and persons. The world around us is not only made up of rocks, plants, animals and humankind. It is inhabited by spirits and gods, but they are not encountered everywhere: it is to their presence or absence that a good part of spatial differentiation is due.

A. P. Elkin[1] furnishes an excellent example of this. The Australian Aborigines live in a very monotonous environment in which they have not undertaken any notable improvement – even if their imprint can be felt in the vegetated landscapes where they have propagated fires. The European colonists had no scruples about displacing the populations that seemed so unrooted: would they not find a similar environment from the one they were forced out of a few dozen or a few hundred kilometres away? To general surprise, these transplantations produced dramatic traumas. Incapable of facing up to disorientation, some Aborigines let themselves die. They had left something irreplaceable in their place of origin: the rocks where the heroic founders of their tribes had chosen to settle, thus linking the significance of the world to a scarp, a gorge or a permanent pool in an arid zone.

Joël Bonnemaison[2] supplied analogous evidence on the universe inhabited by the Tanna people of the Vanuata Archipelago. 'Tanna,' he wrote, 'claims to be the place of origin of all the other islands. Its space is magic, covered by a fabulous network of routes, "sacred places", "tabu ples" in pidgin (figure 7.3). Each one of these sacred

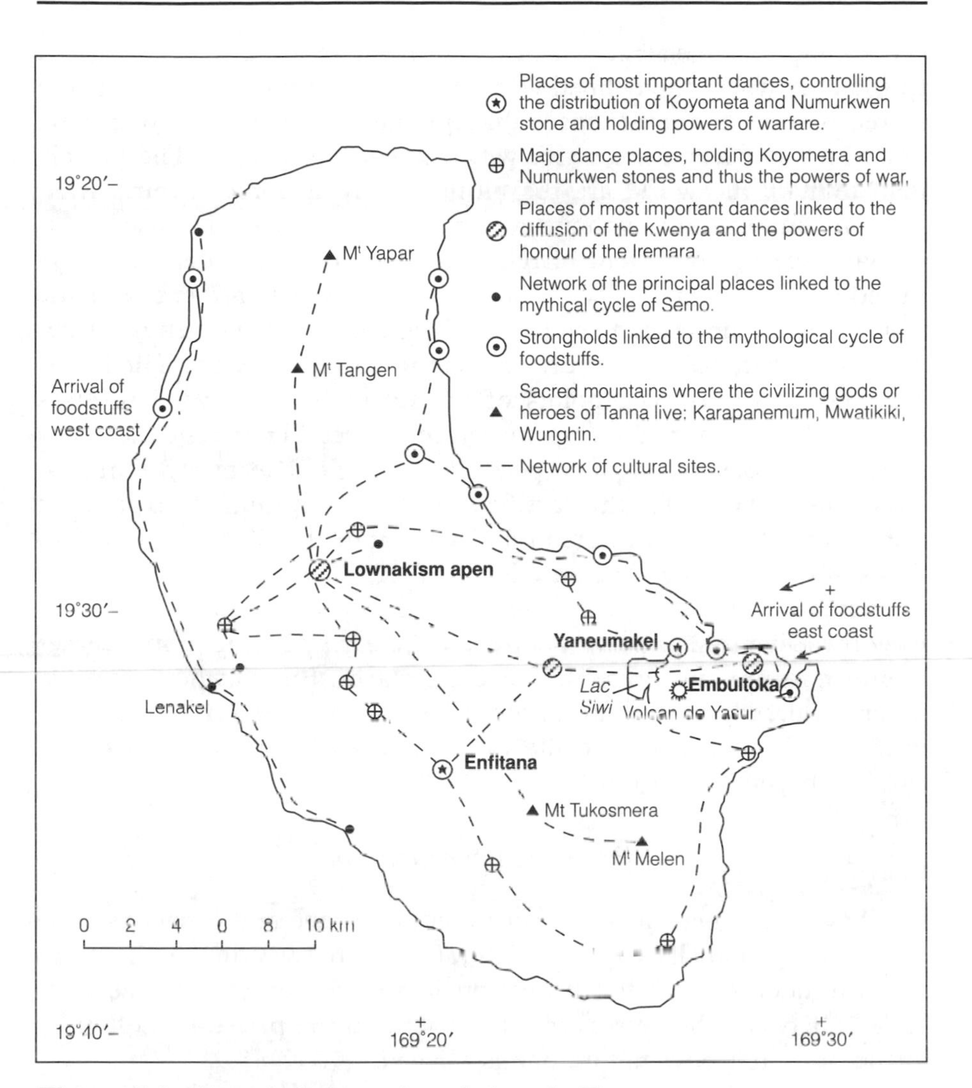

Figure 7.3 Sacred places and mythology in Tanna
Source: J. Bonnemaison, *Tanna: les hommes lieux*, Paris: ORSTOM, 1987, p. 98.

places is associated with a power incarnated in a rock, by a spirit, a hero, a "god" from the traditional pantheon: the sum of these places forms a cosmogony, a sacred geography where the traditional geography of the world is expressed' (Bonnemaison, p. 124). Within this island people, each tribe relates to a dugout canoe, and the men are preceded in the possession of the island by dugout canoes of the gods and spirits who are charged to forge with *chtoniennes* or celestial forces, the alliances necessary for life.

Primitive people see behind beings and objects, genies or gods who are divided between the places where they are manifest and the celestial or the underground world, which are also their dwellings. The geographies of the sacred which result are rich and easy to interpret. In societies where the defined world transcends things and beings, the sacred reposes in another space that the mind is capable of conceiving but which is never coextensive with the earth's surface. The powers that animate the world are the ultimate driving forces behind what happens on earth but without ever inhabiting a terrestrial place.

Many see in the advent of religions or metaphysics depending on the idea of transcendence, the first step in disenchantment with the world and the undefined extension of the profane, which ends up depriving the earth of its flavour. This is too absolute a view. The forces shaping the world need points of impact to be effective. Universes that mirror our own crop up in certain places. Transcendence thus becomes concrete there: God performs prodigious feats and miracles. Catholics gathered for the sacrifice of Holy Communion experience the transubstantiation of bread and wine into the flesh and blood of Christ. Whereas, the Eucharist for the Reformed Church is no more than a commemorative meal. Does that mean that from their point of view the Divinity has permanently left the earth? No: it simply ceases to affirm its presence in certain meeting places. But all who are visited by grace already participate in the glory of the chosen: it is humanity itself that is consecrated to the extent that each one of its members can be chosen by the Lord.

Power and consecration

In all systems where power is religious in essence and is exercised by divine delegation, its leader participates in consecration. It is sometimes difficult to make a distinction in the origins of political institutions between the person who is leader and the priest: it is often by virtue of sacred powers that one is obeyed.

The geography of places where power is exercised and the holders reside, goes back to the idea of sacredness – weaker and less pure than that of the temples, but just as real. The crisis of political systems based on divine law has caused any reference to the afterlife to disappear. Today, political leaders do not think themselves to be of an essence any different from the community of mortals. Their legitimacy is no longer based on transcendence: they no longer derive their power from the Lord. The source of all sovereignty is immediate: it comes from the people on behalf of whom one acts. But since it represents the People, the Power has the right to be respected. Its sacredness lives on, marked by the symbolism focused on it: the governors live in a majestic setting and inspire, according to the neo-classical criteria fashionable at the end of the eighteenth and beginning of the nineteenth centuries, the recreated pomp of Roman monuments.

For the rationalist metaphysicians of the eighteenth century it was the Enlightenment from which the principles that organized the world radiated: everything religious or derived from power ceased to be respectable. It was through the scholars, philosophers and writers that the divinities of the afterlife should be explored. Respect gravitated towards knowledge: academics, universities, museums, are the places where the spirit could blossom, serving as a bridge towards the new 'other world'.

Three centuries ago, criticisms of religious, metaphysical or political forms of transcendence began to multiply. The whole edifice had apparently been destroyed by materialist philosophies of history: for them no other transcendent world existed from which one could overlook earthly realities and which could legitimate the moral authority and political power of clergy or rulers. There is no other truth apart from the physical world. The idea of a better future, the outlines of which could already be guessed at by the great thinkers, nevertheless reintroduced what had just been condemned: those who have the privilege to seize or lead history have thus achieved access to the truth of destinies. This transcendence of an unexpected type gives the prophets of a new era the necessary aura to experience power.

Has consecration disappeared? Let us see. It was never so alive as under the socialist regimes of Eastern Europe. In the places where the great theoreticians of socialism and the heroes of the Revolution lived, their lives were exalted. In Moscow, Lenin's mausoleum was consecrated in the Red Square. Almost everywhere, groups of giant statues recalling the struggle against Nazism animate the square, a little as the calvaries structure the Breton religious space.

Rituals, feast days and the reconsecration of space

Ceremonies and feasts interrupt the flow of time and permit the cosmic time of their origins to break into the world: on the occasion of processions or great collective ceremonies, vast sections of space are suddenly invested with sacredness. On Rogation Day, all the land of a village is commended to God's care. At Christmas or Easter, it is the birth, death and resurrection of Christ that is universally celebrated: the Lord is present in all the places of worship and in families throughout the commemoration.

In modern towns, the rituals have taken other forms. It is no longer the church that summons the gathering and organizes processions. Political parties and trade unions reprove ceremonies, which in their belief belong to another age, but they call their militants and their sympathizers to mass demonstrations. They lead them in a slow stroll, punctuated by revolutionary songs and slogans from one high point to another of the future struggle. They thus rid the Old World of its impurities and incite their followers to new enthusiasms.

The earth's differentiation is not only a material matter. It springs from what each one sweeps up from the world through eyes charged with personal memories and professional experience. It finds, beyond what is provided by the functioning of nature and society, a symbolism which reminds us that the imaginary is an essential dimension of collective life. It interprets the contrasted geography of the profane and the sacred that no civilization can ignore: the rationalism fashionable in the last century made us believe that these were obsolete notions. The slogans have changed but not the realities.

The experience that each person has of life, and the way in which culture develops nature and certain places, thus creates a topology infinitely more varied than a superficial examination reveals. The hurried traveller is only aware of the greater or lesser harmony of forms and landscapes. For the person living there or who has learned to know a *pays*, the emotive power of a locality, the collective memories associated with it, the place they hold in religious life or whatever has replaced it, exude an infinitely stronger and more pregnant image. Without this, one could not understand the role that the territory plays in the formation of individual and collective identities.

7.4 Identity, Territory and Regional Differentiation

The need for identification

Why does the area in which a person evolves hold such a place in their life? It is true that it helps to orientate them, to provide landmarks, to meet partners that they need at an opportune moment, and to know where to relax, laugh and dine without worrying about time passing. But if they only served this purpose, would the types and classifications of space transmitted by society be needed? Certainly not. If regional divisions are so important, it is because they teach us something about our being and our nature.

We cannot avoid asking ourselves who we are, what is the purpose of us being here. We appear on earth by chance and we will leave it one day; these are the only certainties. Our life has no significance unless it is set within a wider destiny beyond the contingencies of individual existence. We feel the need to identify ourselves with collectives that will outlive us. We only exist if we belong to groups which accord us a place; this gives us a sense of security and accomplishment.

Culture is the material by which individuals define themselves and create an identity. Personality is structured at the age of adolescence;

it is the time when the young begin to assume responsibilities. They only succeed if they possess a body of rules to guide their actions. The values held by all are the same everywhere: according to the case, people are all equal or fundamentally unequal. Certain tasks are dignified and others considered servile. The desirable lifestyle is that of the warrior, the shopkeeper, the farmer or pastor. The desirable setting is here the town, there the countryside. The behaviour that distinguishes the members of a group springs from the demands that they have interiorized.

External constraints also play a decisive role; one struggles not to allow the group image to be demeaned; everyone must conform to the accepted model! The members of neighbouring groups form an impression of you and transmit it back to you. Often it is a caricature. But how can one change this spiteful portrayal? This is not achieved by protest; one pretends to accept it; one maintains one's accent, one accentuates one's posture. This is a way of giving positive value to critical judgements; at the same time one subverts the image implicit in the judgement by accentuating the qualities it ignores.

Identity, the sense of belonging and class consciousness

As soon as there exist within a population people who challenge certain attributes, a collectivity is constituted objectively. But as long as they are unaware of what attaches them to this ensemble, their behaviour is not dictated by this belonging. The transition from the collectivity to the class requires information to circulate, for each person to discover what he possesses in common with others, and he must consider it sufficiently important to become concrete, deserve his attention and justify action in common. Therefore dispersion constitutes the first obstacle to the emergence of shared consciousness and identities. The second difficulty results from the impossibility for an individual to exploit by direct contact all the groups he belongs to and which are potential classes.

According to cases and situations, it is a shared faith, favoured recreation, common origins, position in the social hierarchy, to occupation or belonging to the same geographical area, that guides one's preference. The social groups that give birth to feelings of belonging and awareness do not do so in a constant manner nor with a uniformly sustained intensity. In a society with a well-organized economy menaced by nothing, the division of labour and wealth manifests potential cleavages, but the tensions they provoke are seldom dramatized. When a crisis threatens certain people's incomes, or during a phase of political instability when everyone feels that their position could be at risk, attitudes harden: categories become classes and people fight to protect their interests. The whole of social life seems determined by the opposition between those with possessions and the workers, between peasants and their landlords or creditors, and between workers and their bosses. Once the tension is over, oppositions become united and society becomes restructured around other cleavages.

Geographically concentrated groups are among those that lend themselves best to the development of solidarities: problems of communication are reduced to a minimum when people live in the same *pays* and have a thousand opportunities to meet at the market, during a ceremony or at a match. There is no need for the media to disseminate the news needed for awareness. For a long time it was within the framework of geographical classes constituted by local cells, *pays*, the regions or nations, that solidarity movements, self-defence reactions and collective initiatives were normally manifested. The industrial and transport revolution led to a concentration of certain sections of the population. Workers were gathered in industrial areas or large towns; the social struggles, which until then had remained confined to the most important cities, those which enriched business or manufacturing, assumed a national, even an international dimension. The geographical classes, which regions are, appeared less significant.

In the enlarged world that we live in, the horizons of travel, contacts and relations have expanded prodigiously. Often, it is with distant partners, dispersed all over the world, that one has the strongest solidarity and with whom one demonstrates class reactions. This is a second reason for the decline in feelings of attachment to a territorial base.

However, geographical solidarity has not disappeared. Electoral geography demonstrated this throughout the nineteenth century and, frequently, well into the twentieth century. Members of parliament are recruited from the ranks of the well known: they are elected because of their local visibility, because they belong to a great family, because their social success inspires confidence or because their occupation has brought them into contact with everyone. The mass political parties, which try to rally the votes of particular social classes, peasants, workers, salaried staff or managers, take a long time to make inroads outside the very large cities and the cities most affected by industrialization.

Today, in a society where the wages of a high proportion of the population are to be found in the same average bracket, the divisions that are considered significant are not the same. As the old political ideologies lose their attraction, people group around other themes. Religious questions receive more attention than during the two previous generations, but the revival is to the advantage of sects and marginal groups rather than to the highly structured communities of the major religions. The ideologies which served to support national identities are affected by the decline of historical philosophies: the nation state has ceased to appear as a privileged frame within which humanity must fulfil its destiny. The counter-cultures that flourish in this new context do not all adhere to Zen Buddhism or transcendental meditation. The solid popular sources of bygone days also find adherents. They are expressed in local and regional groups. This revives, at least for intellectuals and the newly converted, the charm of provincial life, the lower levels of territorial consciousness, those of the regional scale in particular.

In the world born of industry, what is most willingly condemned is the proliferation of pollution and the often serious damage done to the environment as a result. In response there are demands for a rehabilitation of nature. Immediately spatial groups assume a new significance: one must unite to save the rivers and streams, limit the emission of gasses causing acid rain, avoid the concreting over of the shores and the harmful consequences of mass tourism. The new regional consciousness colours itself green.

Geographical identities: ideological attributes and dimensions

Space is a major factor cementing groups. It gives a very strong sentimental coloration to identities based on territories – the term 'geographical' is still used – and in a secondary role, contributes to encouraging those based on common origins, similar ways of life or shared roles and convictions.

This is what is generally invoked by those for whom the *pays* is the basis of an individual's identity: the environment in which one lives, with its varied relief, its forests, its meadows, its farms hidden by hedges, its villages, supplies landmarks the combination of which is never identical. Is it not because one is a part, like the fields, like the houses or the roads, of such an ensemble, that one possesses something in common and that is contrasted with others? Groups readily insist on the singularity of their country. The links woven between the individual, the community and space are born of the universal experience of the discovery of the world: everyone is attached to the *pays* from one's first steps, from one's mother's first smile, the first games, and later one's first emotional feelings.

What makes one identify with a territory is that the environment is not a neutral framework: each one of us is capable of attaching a personal experience to it, to situate one's family traditions and to discern the group memory. Strong collective attachments are born in this manner, even though no one has invested exactly the same part of himself and does not experience the same affection. An image is sufficient to remind the members of a group of the ensemble to which they belong. The region, as a cultural reality, is based on symbols. A place, a small *pays*, often serves to underpin the consciousness of an entire ensemble: in Franche-Comté, it is the mountain, the dark forests of pine and spruce, and the mountain-dweller, proud, upright and enterprising, which gives the province its tone and defines its identity. The scarp and its vineyards count more for Champagne than the depressing horizons of Dry Champagne, so often decried, or the large limestone plateaux of Lorraine. What would the Provence of Aix be like without Saint-Victoire?* In Alsace, the Vosges foothills and the picturesque little towns nestling amongst the vineyards, for many provide a resumé of the region – but others evoke, in the same way, the soaring cathedral of Strasbourg.

The strongest argument that geographical groups can advance to justify their coherence is that of being a native: people are born in a place and belong to it. Ancestors sprang from the local ashes and earth and returned there at their death, in an endless cycle renewed with each generation. Society is thus like Anteus, the giant son of Poseidon and Gaia, who could only renew his strength by touching the earth: separated from its native soil, it withers away.

Collective memory often retains the memory of wanderings which preceded the settling of the group in the space it now occupies. Being a native is thus impossible. But there is no obstacle to new arrivals forging intimate links with the *pays* where they have settled. The theme of the promised land exists in many mythologies: the founding ancestors came from elsewhere, but they were as if in exile. A setting was destined for them throughout eternity. The Hebrews thus wandered until they found Canaan.

The naturalization of a group can equally result from the sacrifice of the ancestors: when they have fought heroically to defend the land of their *pays* and are buried there, they have made it theirs, and bequeath it as a country to all their descendants. From this springs the significance, for establishing roots, of tombs, cemeteries and funeral rituals. It is important to weep for one's dead every year and to pray at their graves. In some religions, the blood pact linked to the earth is more cruel: humans are sacrificed to root the group to their lands and to mark its frontiers solemnly. What else does the history of the foundation of Rome, the myth of Romulus and Remus tell us? In our civilizations, war memorials recall that the privileged alliance between the group and the territory has not disappeared: religions weaken, but political power saves the ancestor cult by laicizing it.

* The mont Saint-Victoire: a dramatic limestone anticline dominating Aix-en-Provence.

If space only had functional attributes, it would never be more than a neutral framework where existence never knew happenings, major new beginnings, renunciations or revolutions. Because it has seen our childhood unfold, our later portions of our existence, it has transmitted part of the collective memory to us and given it a special accent. Since it is steeped in sacredness, space transforms itself into a part of us: it becomes, to use a current term, a territory.

The hierarchy of identities

Identities interlock and complete each other: those founded on belonging to the small scale birthplace are integrated into larger units, which vary according to circumstances and the partners. When one meets someone from the same *pays*, one states what commune (district) one comes from. In the regional capital, one mentions one's native *département*. In the presence of foreigners, one simply says that one is French. It is during adolescence, when one understands that one must situate oneself within the framework of wider society, that this dialectic of scales is expressed.

How are feelings of geographical identity hierarchized? The *pays* of everyday existence, however badly it is defined, is that of the fundamental community, that in which one plays a full part without it being necessary to talk about it. It is because one is accepted by this base group that one belongs by right to the structures that surround it. When the *pays* has outlines and a content made precise by a name, identification becomes more conscious and aids a definition of contrasts, often slight, but which are strongly felt.

At the medium scale of the region, the different entities that one can distinguish are not all invested with the same emotive charge. To say that one is from the East or the West (of France) resonates just a few common characteristics and cannot create a strong feeling of territorial belonging. Matters are different in the North, which before the Revolution was divided into minor provinces, on the scale of a *pays* more than a region, and which had become aware of its cultural and social identity under this apparently vague geographical label. The Midi expresses a deeply felt sentiment, of a certain *occitane** identity.

Often it is the old provinces that remain endowed with the strongest sentiment. For example, in the East, one feels oneself to be from Champagne, Lorraine, Alsace, Franche-Comté or Burgundy. Everyone belongs to one or other of these territorial communities. Their stability persists even where contacts integrate the *pays* to other ensembles: the south of Franche-Comté is oriented towards Geneva or

* Occitane refers to the language and literature of the *langue d'oc* of Provençal French.

Lyon; but one asserts that one is from Comté. Mâcon feels itself to be Burgundian in spite of the proximity of the capital of the Rhône-Alpes region, Lyon, to which almost all movements take place. It is in such cases that one can measure the advantage, in people's minds, of entities long since recognized and of which history has established the profile, specified the frontiers, revealed the qualities and underlined the weaknesses.

Territorial constructions linked to the sphere of influence of cities certainly count for more in economic life, but they have blurred limits and have lacked the time to penetrate as deeply into collective wisdom, such that people are less moulded, during their youth, to accept them as being fundamental. Born of commercial activity and business, the emotional charge brought by a history of struggles, glory or misery is also lacking.

7.5 The Models of the World, Nature, Countryside and Town

The spirit never remains passive in the face of the world. To the perceived image it is capable of opposing those that it has constructed. To the order of what exists, it juxtaposes that which must exist. It is through these elsewheres thus defined that it is possible to judge the real, to say if what we observe is good or bad. From this starting point, models are constructed that we try to impose on nature and society. Spatial organization does not only originate from the influence of ecological chains or social and economic mechanisms. It also results from the effort people make so that the real world conforms to their norms. Each group possesses techniques which permit it to exploit the environment and to become organized, and models which guide its actions. There is always, in the display afforded by humanized regions, an element of planned development.

The models which have been undertaken are very diverse. We know through the studies of Claude Lévi-Strauss[3] devoted to the Bororos, or through those made by Griaule[4] of the Dogons, that these groups tried to translate, in the organization of their habitat, a vision of the social world and an image of the cosmos. In the Chinese cultural area, geomancy permitted the reconciliation of world order and the practical organization of farmlands and settlements. Paul Wheatley[5] has shown the religious inspiration that presided over the choice of the 'pivot of the four corners' in the imperial cities of the Chinese high antiquity. Similar preoccupations were present in Mesopotamia as well as in the great pre-Columban civilizations. Joseph Rykwert[6] has been able to detect, underlying the geometric plans of Roman surveyors, the same religious preoccupations.

Since the Renaissance, it is often in the direction of aesthetic doctrines that one must turn to understand the mental maps which influenced the tracing of towns or the layout of gardens. But it is the concern to eliminate wrongs linked to the inequality of status and wealth circulated since Thomas More in all the Utopias that influenced, and continue to influence, the ways in which we organize space.

Conclusion

As we have seen, regional entities can be defined in many ways. What the cultural point of view brings to regional analysis is the idea that the spatial divisions proceeded to by societies are frameworks which do not leave their members indifferent. Since in the past territories generally differed in terms of their speech and by the techniques that had been transmitted to them, since they served as a framework for fixed existences with a thousand real hardships and since they are the last resting place of parents and friends, they did not remain, and will never remain, neutral. Regional spaces owe to this their very development in the first place or in a secondary capacity, the foundations on which social identities are defined. They stimulate collective consciousness, they introduce solidarity in people's lives, give birth to reactions, motivate strategies and sometimes serve as a basis for collective action. They are not simply a convenient framework for analysing geographical reality. They constitute one of the echelons of all the social dynamics which shape the face of the earth.

The increase in mobility and the extension of holidays are translated by a rapid shake-up in feelings of belonging. They lead to a banalization of standards of living to the extent that ways of life are tending to conform everywhere to the same models. But, simultaneously, regional identities sometimes appear as substitutes for feelings that were stronger in the past, but which are victims of a more severe erosion. Nationalism was reinforced by the idea of a construction to build in common, of a destiny to achieve to confer on its group the benefits of all the advantages of progress. In proportion to the generalization of economic development, nations lost the privileges gained from being pioneers. All the nations of Europe enjoy more or less the same social rights: this removes the advantage of being French rather than German or Italian.

The crisis in feelings of belonging to a nation thus makes regional identity fashionable again. It is however different from that of the past. In societies where mobility was weak, belonging was experienced as an imposed fact, as an element of destiny which must be assumed but which was not chosen. Today, the fashion for regionalism leads people to identify with such or such an ensemble because it pleases them, because it offers agreeable landscapes, a clement sky, well-serviced towns, or because it was celebrated in literature, poetry or the cinema.

Our world is confronted by an upsurge of the irrational: western states try to remain above particularisms and to extinguish quarrels and local vendettas. They do this in the name of progress. The crisis in the political philosophies on which the nation was built has made types of identity fashionable again that previously were considered obsolete. People are willing to fight to preserve minor differences at a time when life is impossible without exchanges extended across the whole universe! It is this that makes both passionate and tragic the study of regionalist ideologies and the passion the world undertakes in the name of territory at a time when everything is tending to be mobile.

Further Reading

Buttimer, A., Grasping the dynamism of the life-world, *Annals of the Association of American Geographers*, vol. 66, 1976, pp. 277–92.

Dirven, E., Groenewagen, J. and Hoof, S. van (eds), Stuck in the region? Changing scales for regional identities, *Netherlands Geographical Studies*, 1993, 117 pp.

Duncan, J. and Ley, D. (eds), *Place, Culture, Representation*, London: Routledge, 1993, 341 pp.

Jackson, P. and Penrose, J. (eds), *Constructions of Race, Place and Nation*, London: University College of London Press, 1993, 216 pp.

——, *Maps of Meaning: An Introduction to Cultural Geography*, London: Unwin Hyman, 2nd edn, 1994, 213 pp.

Keith, M. and Pile, S. (eds), *Place and the Politics of Identity*, London: Routledge, 1993, 235 pp.

King, A. (ed.), *Culture, Globalization and the World System: Contemporary Conditions for the Representation of Identity*, London: Macmillan, 1991, 184 pp.

McDowell, L., Towards an understanding of the gender division in urban space, *Environment and Planning, D: Society and Space*, vol. 1, 1983, pp. 59–72.

—— and Sharp, J. (eds), *Space, Gender and Knowledge*, London: Arnold, 1997, 468 pp.

Olavig, K., *Nature's Ideological Landscapes*, London: Allen and Unwin, 1984, 115 pp.

Rose, G., *Feminism and Geography: The Limits of Geographical Knowledge*, 1993.

Seaman, D. and Mugerauer, R. (eds), *Dwelling, Place and Environment: Towards a Phenomenology of Person and World*, Dordrecht: Martinus Nijhoff, 1985.

Short, J. R., *Imagined Country: Society, Culture and Environment*, London: Routledge, 1991, 253 pp.

Smith, D., *Geography and Social Justice*, Oxford: Blackwell, 1994, 344 pp.

8

The Region and Political Life

The structuring and regional differentiation of the earth do not translate only the influence of natural forces, the influence of social mechanisms and the exploitation by human groups of the zones where they live and exercise their activities. Public administration contributes in large measure to the hierarchical ordering of urban centres and they are responsible for the only divisions with linear boundaries. It permits those in power to control space and the people implanted within it. The authorities thus dispose of the means to implement planned development and to intervene in a territory's organization. We will now explore these aspects of the shaping of regions. To elucidate this we need a few ideas on the relationships between the political system and the areas that it structures, and on the way in which the services it provides are integrated into the territorial structures of the public at large.

In many societies, the necessary harmonization of individual decisions takes place within structures of a familial type, or results from the influence of associations, castes or orders. The political dimension of collective life is not expressed in formal institutions. In this case there are no functional spatial organizations exceeding the scale of the *pays* or region. All that is encountered are homogeneous ensembles linked to the strength of natural constraints, to the origins of the peopling and to the conditions of cultural diffusion. Obviously, the developments discussed below are not concerned with those societies.

8.1 Civil Society, Political Systems and the Territorialization of Power

Public services, civil society and political system

Human societies need security: they must be defended against external threats and prevent harm from those who attack private persons or the basis of collective life. The political system permits this contradiction to be overcome: it provides citizens with indispensable services by taxing part of their income. Thus it is responsible for defence, diplomacy, the police, justice and the establishing of laws and regulations indispensable to the smooth running of private business.

Citizens are so well aware of the importance of health or education services that they are willing to pay a high price. Unfortunately, low income groups have not the means to act in this way and consume beyond their means. This explains some of the recent extension in the field of public intervention. The Welfare State taxes the rich and transfers the income thus derived to the poor. It controls the provision of health and education services itself so as to guarantee a fair distribution. Its domain is no longer limited to security and public order; it is expanded to include social justice. This invokes the principle of solidarity and implies transfers, which always have important geographical effects.

Public facilities now offer the services considered indispensable for everyone. It is accepted that all have the right to be served by road, to have mail delivered regularly, to have a telephone line, to be connected to the electric grid and to receive treated water. To limit the amount of pollution entering the water-table, there is an obligation upon all to be connected to the sewerage system that the authority has established. It was long thought that these facilities were a privilege and that authorities were not obliged to provide a universal service. In France, if someone wished to benefit from the roads that the newly created *Ponts et Chaussées* administration had built at great expense, they had only to settle in districts served by the network of *Routes Royales*! A change in attitude became apparent when a judicial framework was established to provide for the financing of local roads under the July Monarchy.* The idea that all had a right to an electricity supply was asserted in the interwar period. For drinking water and sewerage provision, the stage was not reached until the 1950s. The same evolution has scarcely been attained for the provision of telephone facilities or, in rural areas, for the collection and treatment of

* The monarchy under the King Louis-Philippe, abolished by the Republican Revolution of 1848.

rubbish. It is evident that the range of facilities paid for by the public sector has enlarged.

To understand political life it is convenient to distinguish between civil society, which is everything in everyday life that concerns private relationships and functions without public regulation, and the political system, which provides the distribution of public services and resolves conflicts which otherwise would be insoluble.

Economic mechanisms at work in civil society are not always perfect. The political system often tries to control and improve them. When the results obtained in this way are considered unsatisfactory, another stage is often crossed. Nationalization transfers to the state sector tasks previously assumed by the private sector. Socialist regimes have gone further in this direction, but the results obtained have been mediocre. Interventionism has become unfashionable and everywhere there is an attempt to give back to civil society the initiative and responsibility that was taken from it.

Relations take place between the sphere of the state and that woven by the links uniting private persons or enterprises. Those in power need to be informed of what is happening within the territory occupied by civil society and to apply measures there.

The architecture of links between civil society and political systems

The modern state in essence is unitary. Its role is to eliminate the diversity of situations engendered by the events during the peopling, the transformations of history, the divergences of local traditions and the proliferation of ideologies. The political system tries to standardize the legal basis on which civil society rests: citizens enjoy the same rights everywhere, have the same prerogatives and are subject to the same obligations. The national territory is born by denying the uncontrolled and irrational forms of diversity. But this negation is only possible through an effective control over space: it is because the objective of the political system is to unify the judicial framework of existence and thus to facilitate trade, the movement of goods and personal travel, that the state must maintain its control over local or regional units. It can only achieve the uniformity it wishes to establish by action based on territorialization.

The public services provided by the political system have no value to the citizenry unless they are accessible. It is not enough, for the citizens to sleep peacefully, to create police forces and courts in the capital. The control of deviant behaviour is only possible by the division of space into administrative units with clear boundaries. Officers are appointed there, who put the dangerous individuals under permanent surveillance, visit flashpoints frequently and receive intelligence from widespread informers. When a conflict breaks out between individuals which cannot be resolved out of court, it is for the judiciary to explain to each one their rights and to apply them. We cannot have recourse to the law unless tribunals exist nearby and are easily accessible. The creation of a space, where the differentiations created by physical factors and history have been erased in favour of uniform control therefore demand a strict connection between civil society and the political system.

8.2 The Functions of the Political System and Administrative Divisions

The political system and data gathering

A political system can only work usefully when it is informed. To achieve this, two strategies are employed. They rest on the construction of systems of representation and on the creation of an administrative structure.

The best way to have reliable information concerning problems arising in a territory is to consult the population. All political systems have recourse to this: the monarchies of the *ancien régime*, for example,

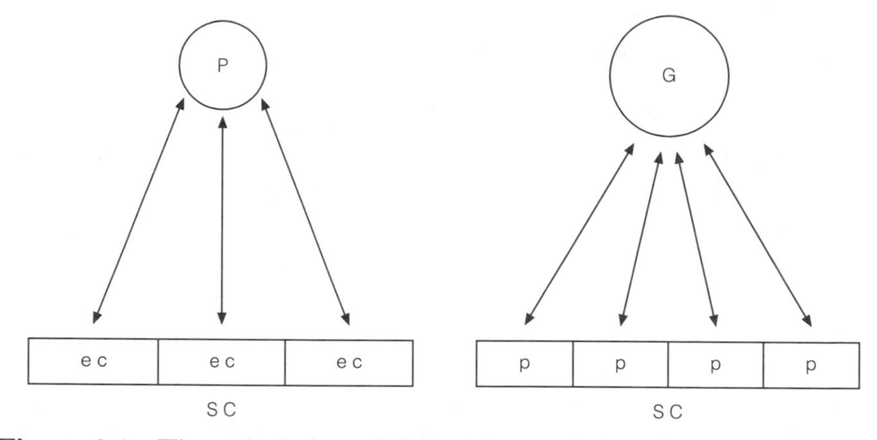

Figure 8.1 The articulation of civil society and the political system

1 The political world (P) receives its legitimacy from the civil society (SC) on the occasion of elections carried out in electoral constituencies (ec) to name members of Parliament (P).

2 The government (G) receives information about civil society (SC) through *préfectures* (p) and transmits to them orders intended to enforce laws and regulations. Electoral districts and administrative divisions often do not coincide.

convened the *Etats Généraux.** With modern conceptions of the state, mass representation has transformed and enlarged its roles. It supplies a legitimate basis for the exercise of power and guarantees the upward transmission of the information necessary for the drafting of public decisions.

It is not necessary for a country to be divided into electoral districts so that all currents of opinion can be expressed: the electoral systems of the Netherlands and Israel prove this. But it is in small-sized states that the data collection indispensable to the government is easy. In larger states, consultations are organized on the basis of electoral constituencies. It is the only way to be made aware of all the regional problems. The fine detail of currents of opinion is lost in the global representation, but one is better able to measure the conflicts developing in one part or another of the national space.

There are many ways of defining electoral constituencies. It can be useful to make them coincide with administrative units: as these often have a similar size, it is the solution that has been adopted for a long time by the majority of states. But their densities vary from region to region. The *cercles*, *arrondissements*, or *départements* have not the same population size, which creates an injustice: not all the elected members represent the same number of electors. The demand for equality has increased in our lifetime. We define ad hoc constituencies: they should have roughly the same number of electors enrolled (the maximum difference must be under 10 per cent in some countries and 5 per cent in others). The operation provokes endless debates since numerous solutions are possible.

The second way of ensuring that information is transmitted upwards is to take the administrative approach: the appointed civil servants, each in their particular field, take the pulse of the population. Space is divided into administrative units according to a mesh which varies in relation to the nature of the environment and population density, that is according to the difficulty of observation. The coverage must be total and the network of the divisions sufficiently fine so as to allow nothing to escape. The base units coincide with the perimeters within which daily life takes place, the equivalent of the French commune.

To ensure the centralization of the data captured, a hierarchical structure is necessary. Data require to be treated before they are dispatched: they must be sorted, controlled, summarized and only what seems to be pertinent circulated upwards from the small to the large towns and then to the capital: the pyramid of administrative towns can be independent of that involved in the provision of commercial services or can coincide with it.

* Representatives of the province convened by the monarchs in pre-Revolutionary France to debate important affairs of state.

The provision of public services

The political system must be rooted territorially to achieve the provision of public services. These are subjected to similar threshold limits as those encountered by other services. They differ in only one respect: a client is not obliged to patronize the nearest shop whereas the subject in an administrative sense finds himself assigned to a centre that he depends on as a function of his domicile.

The coincidence between the administrative structuring of space and that which derives from relations within civil society is practically total at the lower levels of the hierarchy. When the establishment of a commercial economy dates far back into the past, the administrative divisions are generally moulded on the hierarchy of pre-existing villages, small and larger towns. Conversely, when a state decides to supply the distribution of public services in areas which until then were private and lived in an enclosed world, without towns, the new administrative centre locates the businesses and private services which are not slow to proliferate.

The services supplied by the state are very diverse and do not all have the same threshold limit. Thus each administration tends to establish a mesh which suits its own needs: one mesh for the postal service, one for schools or for dispensaries, one for the registry of births, deaths and marriages, one for the taxation office, one for property titles, one for weights and measures, etc. (see figure 8.2). The solution is advantageous in terms of public finance since it permits the number of civil servants to be reduced to a minimum, but it does not suit the citizens as it complicates their movements. Firms are often even more inconvenienced: the administrative problems that they have to resolve are complex and do not fall within the competence of the lower levels of the hierarchy.

The administrative system installs public facilities. It has responsibility for the construction and maintenance of means of communication and telecommunication systems, or at least for their control. Alongside linear infrastructure, point-based facilities must be created: schools, high schools, universities, clinics, hospitals, asylums, but also prisons, barracks, etc. Some choices are essentially dictated by national criteria (this is the case of military or naval facilities) and others are determined by the service to a population the density of which is uneven and diverse. The administrations, para-public services or the private firms charged with creating, operating and maintaining these services under the responsibility of the state need to be structured territorially. The Highways Department, for example, must be present very low down in the hierarchy to provide the daily maintenance of roads. At a much higher level, it must have more elaborate services to carry

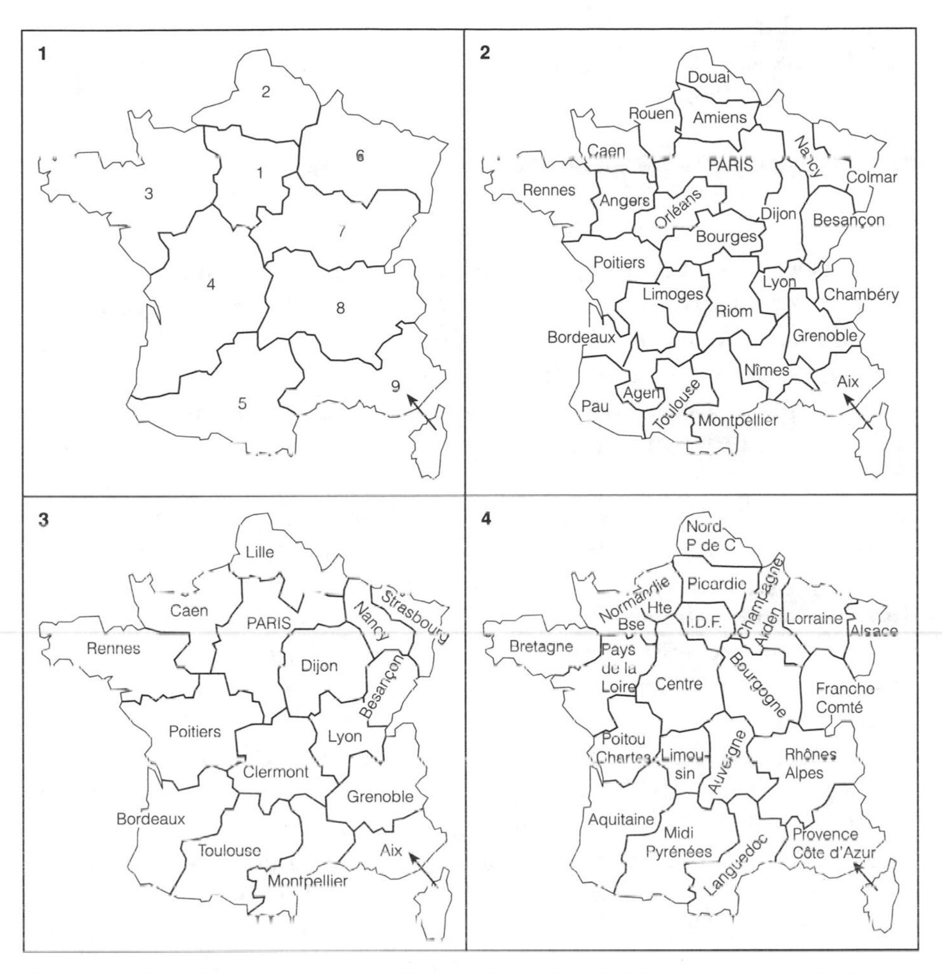

Figure 8.2 The multiplicity of administrative divisions in France before the 1962 reform

In 1960, the 9 military regions (1), the 27 judiciary districts (2) and the 16 university academies (3) retained their traditional units. It was on the regional planning districts (4), which numbered 21, established in 1955, that the standardization of boundaries took place subsequently.

out studies necessitated by the opening of new links, to direct their construction, or to supervise it if the task is contracted to a private company, and to assume financial responsibility. It is also at this level that contacts with companies and other administrations are established.

Schools, colleges and high schools must be located as a function of the population and so as to avoid long journeys for children or adolescents. All the population must benefit from accessibility on as equal a basis as possible. The alignment of a road must meet local, regional or national requirements, which implies collaboration between communal, *département*, and regional authorities and the carrying out of national objectives. The administrators charged with the installation of facilities have a difficult task for they must take into account interests that are often contradictory.

The function of stimulation and animation

One has long been tempted to consider that economic life and cultural animation were the business of the private sector and associations. This limited the action of the state at a territorial level to the installation of facilities and the distribution of public services. The lack of coordination between these initiatives, which often spread across several administrative structures, was scarcely favourable to the general interest.

Today, attitudes are very different. More attention is paid to the role of the institutional environment in the fortune of enterprises. The first stages of development of a firm in a new industrial site are always difficult: some infrastructure is lacking; the public services have not the means to supply essential grants or to provide them at the required frequency. If it is wished that the transplant should succeed, a coordinated action by public administration is indispensable: it progresses through the creation of industrial estates, the construction of turn-key factories, the installation of networks, the extension of public transport and the building of housing for the newly employed labour force. The idea that all the regions should have equal opportunities encouraged, with the *dirigiste* philosophies of the 1950s and 1960s, the proliferation of interventions of this kind. Even with this public sector support, the costs that a firm must accept are so high that they weigh heavily on the balance sheet during the first years. Fiscal concessions are extremely useful to navigate this difficult period.

In the past, the facilities required by industrial firms and services were less complex; the hurdle of the first years was easier to pass. Conditions have changed. Initiatives only succeed where they take place in an environment that facilitates their task.

The crisis which, since 1973, struck the old industrial regions suffering increasingly keen competition from newly industrialized countries, has accentuated this evolution. No longer can one count, to promote employment in a region, on the creation of factories manufacturing mass consumer goods with unskilled labour. Only high technology activities can withstand competition. To guarantee the success of new companies in these sectors, many conditions must be brought together. As in the past, it is necessary to facilitate the acquisition of facilities and sites, housing for the workforce and the provision of indispensable infrastructure. But the qualifications have changed: the staff are better trained, professional workers, engineers and managers are initiated into modern management methods. Personnel wish to find, away from their workplace, places to relax and to play sports. They need an active intellectual life. Economic development therefore implies coordinating action over a much wider area than twenty years ago. Fashionable liberalism would have us believe that entrepreneurs desire only one thing: to see the state disengage from the economic field. This is false: businessmen expect that their task will be facilitated by creating the environment they need. The need for well-located public facilities, developed in a coordinated manner, has never been stronger.

In a society of extended education, where the working week is shorter and where retirement is earlier, recreation and cultural life become important. It is not simply to satisfy company managers that initiatives must be taken and coordinated in this field, it is to respond to everyone's aspirations. Interest in the national heritage, the concern to display monuments, the wish to avoid the disappearance of vernacular architecture and interest in popular art and culture, imply the need for a cultural policy adapted to each locality, to each *pays*. It aims, in certain areas, to maintain the life of regional languages and dialects.

The systematic recourse to fertilizers and pesticides, the generalization of car ownership and the rapid increase of consumption, especially of energy, have made human impact on the environment greater. Agricultural recession is evident in marginal regions. Environmental management requires measures that are only effective if they are applied according to an integrated plan.

The territorial actions undertaken by a political system respond to very diverse exigencies. Divisions established for elections vary without respect for the demographic evolution. Those used for the distribution of public utilities are managed according to purely technical criteria and, without too much inconvenience, can be incorporated into ad hoc territorial hierarchies. More complex territorial considerations must be taken into account in the case of incentives, support and development that modern life has multiplied. The administrative system in these circumstances cannot ignore the territorial structures of civil society. In some cases, the regional level plays a decisive role.

8.3 The Political Territorial Framework

The place of the region in political life depends to a large extent on the style of the political framework chosen. Three main options are offered in this respect and are often combined.

Direct administration and the role of civil servants

The first way of controlling the national space consists, as far as the government is concerned, of applying the measures taken by civil servants that it employs, remunerates, who receive orders from it and who are answerable for their actions.

In a system based on a civil service, the state is omnipotent in imposing the territorial division that suits it, but it has an interest in drawing at the lowest administrative level on the cells which define people's daily lives, and to locate its services in centres animated by business and inter-relationships. However, it is not mandatory to reproduce the hierarchy of its services exactly on that which results from the functioning of civil society. Each service tends to develop a spatial organization best suited to the mission.

The risk in all centralized administrative systems is to give too much power to local representatives of the state: the temptation to abuse it is great. Civil servants no longer conform to orders given from above, they retain, to reinforce their position, part of the information that they have gathered, commit abuses of authority and become local tyrants. The manoeuvre is easier as financial centralization is defective and as the mobility of the top civil service diminishes. In societies where transport was difficult and communication slow, there was a real menace of seeing the hierarchy of officials mutate into a feudal system. This kind of deviation is no longer possible: local government officials no longer dispose of taxes raised locally; the means that they deploy are allocated to them by central government which carefully controls their use. The practice has also become common of rotating the posting of *préfets*.* The risks of collusion between local society and those charged with making it respect the law is thus averted.

The application of measures decided in the capital always elicits resistance. This can become serious in centralized systems where there is no local mechanism to resolve conflicts. Above all, the state fears the rebellion of entire fractions of the territory. They mistrust local public opinions and whatever can strengthen them. Feelings of belonging to a region upset them. This is why they define divisions smaller than the regions that are alive in the collective consciousness. They posses too much coherence, which makes them difficult to manage when opposition develops. The wise course, therefore, is to

* The *préfet* is the head administrator of the *département*, appointed by and directly responsible to central government.

subdivide traditional territorial groupings with a strong individuality: this is what the *constituants** did at the beginning of the Revolution by replacing the provinces by *départements*.

The problems posed by centralized systems of territorial management also result from the slowness of the process of decision making that they establish. The arbitration preceding a final decision takes place at the top level. It takes place in a calmer atmosphere than locally. Interests that have been expressed are taken into account, but the indirect impact of decisions on development possibilities and on the environmental balance are not calculated. The administrative solution is less appropriate now, which was not the case when the objectives of stimulating and animating development were a lesser political preoccupation.

The system has it advantages. It permits the deployment of a technically qualified personnel – civil servants recruited by rigorous criteria – but it is expensive. It weighs very heavily on the budgets of emerging states in newly developing countries. Here, salaries are often not paid regularly: officials are then tempted to sell their services directly, which opens the door to widespread corruption. The state loses its grip on the territory it must control and it is its agents who set a bad example and provoke demoralization in the whole body public.

With the enlarging of the state's responsibility over the last half century, the number of dossiers to process and decisions to be taken has grown so much that the risk of choking has become evident. Delays extend because offices are overstretched. In the former Soviet countries, where centralization also concerned decisions on production, the consequences quickly become catastrophic: the whole system was threatened with paralysis, and the measures taken to facilitate the centralization of data – the construction of giant units for example – had harmful effects which were more serious than the efficiency gains that were initially achieved.

Everywhere that a centralized system of territorial management is persisted with through an administration entrusted to civil servants, the tendency is for centres of decision making to be located closer to the places to which decisions are applied; in technical terms we refer to decentralization. An increasingly important proportion of choices are made without reference to the capital. The system gains in effectiveness but without the flexibility and adaptability that other solutions offer.

* The *constituants* were members of the *Constituante*, the precursor of the *Assemblée Nationale*, the post-revolutionary elected parliament.

Decentralized territorial administration: territorial groupings, the law and justice

Another way exists for states to organize their administrative systems. In place of the application of measures by civil servants of the central government, they can be devolved to territorial bodies. The danger of fragmentation inherent in such a solution is compounded when two conditions coincide:

1. the law must be drafted in as clear and constraining a way as possible, and without the slightest reservation permitting its application to be restricted;
2. the citizens must have the right to challenge the administration responsible for applying the law in tribunal courts.

The threat carried by the recourse to justice forces local administration to respect legal provisions. In these conditions, decentralized administration does not lead to fragmentation and offers positive advantages. It demonstrates economy: everything possible is done to avoid paying a plethora of workers: the control over work and the pressure to limit waste are stronger than when management is remote. On the other hand, it is to be feared that the positions will be held by incompetent officers because they are underpaid, or are recruited according to political or ideological criteria from supporters or family members of the local political leaders. The risk of misappropriation of public funds is real and can only be limited by a strict audit of contracts and expenses committed.

Local politicians know that they can be dragged before the courts if they do not apply the law or if they apply it in a partisan manner. They risk very heavy penalties if they use public funds to serve their interests. This is sufficient to remove the most blatant forms of local tyranny, without always eliminating the danger of it.

The essential advantage of decentralized administration is that it is easier to take into account local or regional information at the stage when decisions are being prepared. They are better adapted to their context: it is easy, *in situ*, to understand that a better transport infrastructure and an improved quality of life will avoid the exodus of qualified staff and the flight of dynamic firms and that they will perhaps lead to valuable inward investment in plants. When a property development project threatens a church or fine mansion, the mounting protests of associations to protect the heritage cannot be ignored. The ecology movement closely monitors the way in which waste is treated, burned, buried or dispersed. It is ready to denounce firms contributing to pollution, even if the power of the latter ensures that they will receive support in high places.

The federal solution

Territorial authorities often enjoy a greater autonomy. Their role is not restricted to applying laws and rulings promulgated by the central power. They construct a judicial framework adapted to the aspirations, customs and problems of the local society. The federal system thus introduces a considerable latitude between the territories making up the national chequerboard.

The risk of fragmentation is still more obvious than in the case of decentralized local authorities. The number of entities that enjoy a real autonomy is not infinite. The federal principle is generally applied to the upper levels of the hierarchy; it concerns the major regions or powerful urban metropolises. Exceptions still exist: the cantons of the Swiss Confederation are small scale: the most extensive, Berne, is smaller than the majority of French *départements*; those of the original Switzerland, or the urban cantons of Geneva or Basle are minute.

The federal formula is especially adapted to the preservation of religious, ethnic or linguistic diversity. When all the components of the national territory are on the same judicial footing it is rare that one of them can be capable of imposing its culture on the others. This is illustrated by Switzerland or Germany. Countries with great cultural diversity, which accept rather than deny it, willingly adopt a federal structure, as is demonstrated by India or Nigeria. The formula also suits very extensive states where centralization of regulations would impose costly back and forth movements: this is what justifies federalism in Australia, Brazil, Canada or America. It is possible to try to overcome the problem of a large expanse without recourse to federalism: this was the solution adopted by the Soviet Union and also adopted by China.

Federal structures divide the functions of government between several capitals: this is expensive and in the short term may hinder global development, but it increases the number of environments with diversified social structures where all professional activities are represented. At a time when companies have become very sensitive to quality of life, it is an advantage. The distribution of growth is less unequal and all the parts of the national territory benefit.

Mixed solutions

The majority of territorial administrations associate, in varying proportions, the principles the logic of which has been outlined above. The reason is firstly financial: the establishment of a centralized administration is so expensive that generally one avoids going too far in that direction: below a certain level the role is provided by territorial collectivities, in France by the communes. Their mayors, according to the *Larousse Encyclopedia*, have a double function: 'they represent the communes and prepare and execute in this role the decisions of the municipal councils; they are the agents of the central power and are charged with overseeing the publication and execution of official laws and regulations, the maintenance of order, etc.'.

These are the functional reasons why two principles are often employed at the same level. In a decentralized or in a federal system, the citizens must not relax their vigilance when their rights are at stake, but they are not concerned to defend the interests of the national community. The latter must make sure that long-distance links are guaranteed, that the army has access to sites and bases necessary for the national defence and that the educational system furnishes the necessary abilities in the face of international competition. To achieve these objectives the state needs a centralized administrative structure. It is concerned above all with the services which, like the army, do not impinge directly on the structuring of space, but nevertheless have a territorial aspect. The administrative mesh is thus double: it includes territorial collectives and services directly managed by ministries that oversee the enforcement of measures in the national interest. This is the situation in France. Since the Revolution, the *département* has been the major administrative area of the system, but it is also, from its origin, a territorial collective with an elected assembly, the General Council.

Even where territorial collectives are entrusted with applying the law, and even if a federal structure exists, there is sometimes a withdrawal before the power represented by certain units, especially the capital city and metropolitan regions. Washington is located in a federal district controlled by Congress. In the case of London, Mrs Thatcher dissolved the Greater London Council, which opposed her policies. In France, Paris has only had an elected mayor since 1977 and the government is currently trying to limit the powers of the Île-de-France region by reminding it that the central state has more rights and obligations there since it encircles the capital. In Catalonia, the Autonomous Region is still opposed to the constitution of an authority controlling Greater Barcelona, since it would concentrate three-quarters of the population of the province and would dispose of greater financial resources than those of the Catalonian government.

We see then that the place given to the region in the spatial organization is a function of the solution chosen to implement territorial aspects of national policy: the regional level tends to be reduced to insignificance or ignored when direct administration is chosen. A greater importance is given to it if the laws voted by the legislative body and the policies established by the government are delegated to the territorial collectives. In federal systems, the units are often very large and exceed the normal dimensions of regions as conceived by geographers.

In the course of the last thirty years, two factors have generally encouraged a more important place to territorial collectives and gravitation towards federal systems:

1. the choking of central services due to the fields in which public power intervenes;
2. the need to provide a better coordination of action at local level if it is wished to favour the creation of local enterprises, monitor the quality of life and preserve the cultural and natural heritage.

8.4 Regional Actors and the Functioning of Administrative Systems

The space over which the state exercises power is not empty. To maintain control over the society living there, it needs relays and intermediaries. It is insufficient to take advantage of the transport and communication infrastructure and the central places that civil societies have established and developed. It must be supported by intermediary bodies, that is, the groups that structure civil society, and enter a dialogue with the leaders who have arisen from them.

The dialogue between leaders and civil servants, and the real-world functioning of centralized systems

The role of civil servants varies considerably according to the model chosen to apply national policy territorially. It is dominant when the administrative approach is favoured. It is very modest when the territorial collectives are entrusted essentially with the task. An autocratic style is apparent on the part of representatives of central power when it lacks any possibility of permanently controlling their actions and has to trust them. Rome experienced flamboyant proconsuls. At their origin, the European colonies were often governed by strong personalities.

In modern times, a low profile is imposed on the functioning of administrations. Civil servants understand that their mission is not to take a personal line and to make their mark on the region they work in by practising a policy differing from those prevailing elsewhere. The law is the same for everyone: state representatives never speak on their own behalf and know that they can be disavowed at a higher level; it is the system of indirect distant power.

Applied with rigour, central direction by a corps of civil servants is blind: it ignores the diversity of local conditions and leads to conflicts which could have been avoided with a little flexibility. One knows only too well that in highly developed administrative systems it is rare for decisions to be taken without delays. Dossiers shuttle back and forth between the capital and province. Officially, the green light is given from the top, but the decision would not be possible there without information. The local representatives of power furnish it: their role is to evaluate the forseeable effects of measures they are asked to apply and to warn the central power of the reactions that they risk provoking. They need local suppliers of information. As they are obliged to be neutral, they do not have a free choice of their interlocutors. The leading dignitaries are the only ones they can approach without being reproached: locally elected officials and representatives of constituted bodies, presidents of chambers of commerce, of trades or agriculture, major company directors or trade union leaders. The administrative system is based on connected dialogues at the local level.

The civil servant tries to make the intentions of the capital accessible to his interlocutors, to explain the instructions given to him and to guess the reactions that will occur at the top in relation to the evolution of the local situation. It is this that confers power on him. By repeatedly recalling that he is worth nothing, that he does not take decisions, that everything is done at the top level but that he is the obligatory intermediary, he forces everyone to confide in him. He is the only one to know what is being discussed in the capital. Thanks to this he can give advice which is generally appreciated and followed and permits him to have a decisive influence over decisions. To all appearances, he never decides anything; he depends totally upon Paris. In reality, it is through him that local actors know what is afoot in the ministries, and what heads of sections, the cabinet office and the minister, know of provincial reactions. The remoteness of the seat of power gives the civil servant a decisive role. He is simply a drive belt, but the outcome of a discussion depends on what he transmits and the way he presents it. This type of relationship has been best studied in France but it exists in similar forms in all centralized nations.

The style of territorial administration owes to this its generally muffled character. One does not make a fine career as a *préfet* by being unorthodox, by being headstrong, by being abrupt over matters or with people. One must listen, wait, and let solutions mature which then always appear to have come from elsewhere but in which one has played an important part.

The dignitaries displeased by proposed solutions do not remain inactive. They warn deputies, senators and all the pressure groups that they can convince so that a re-examination can take place in the capital. They cannot fight on equal terms with the administration except by having in turn their correspondents who inform them about what is in the offing, and display a capability to establish a compromise by intervening at top level. This is what takes place in the French political landscape due to the power of these major personalities of local life constituted by deputy-mayors or senator-mayors* of important cities.

To achieve a compromise generally accepted by the population, the central tier of government functions like an appeal court. This assumes that the wording of a decision leaves a certain latitude for interpretation: in this way, arrangements can be elaborated to avoid local upheaval. This leads often to paradoxical effects in the central administration; local particularities can often be maintained for a long time under the apparently uniform cloak of national law.

The system forbids civil servants from abusing their authority. It gives a considerable influence to local dignitaries but denies them direct use of power. This does not prevent them from constituting client supporters but limits the possibility of rewarding their fidelity by distribution of favours or posts.

Dignitaries and local public opinion

The second characteristic of regional political life is the position held by leading dignitaries and local opinion movements. The former are present in centralized political systems but do not have such clear clout as in territorial collectivities, whether or not they are autonomous within a federal system. The major ideological currents have less hold, at this scale, than in the nation as a whole. One can make a national career by championing great causes, by defending the country's integrity, by fighting for the values on which it is built, by opposing the exploitation of the poor by the rich. To succeed in a smaller scene, one must above all be well known.

The dignitary is a person who has arrived and as a result has supporters: they are ready to follow him, vote for him and trust him. It is simply expected that he will defend local interests and he should not forget those who support him. The rules of democracy are beautiful: all people are born equal before the law. But when one has had experience of local life, one knows that certain individuals are more equal than others, and that to succeed, it is useful to count on the recommendations of a well-placed person. The importance of clientelism varies between regions and temperaments, but it always tends to be stronger on the local or regional scene than on the national scene.

The dignitary is a public persona. He is ready to receive and listen to all who wish to relate their problems to him, tell him what wrongs

* Under the French system of *double mandat*, an elected parliamentary deputy or senator can also hold office as a mayor.

they suffer and ask for his support. He answers telephone calls at all hours. He is present at local association meetings the support of which he counts on. He would not be able to succeed without the backing of public opinion.

Ideological debates seem dead at this level. The local or regional press depends greatly on advertising and avoids taking sides. Before the post-war concentration of the press, there were often several rival papers on sale in the same area, which allowed each to express clear political preferences and to have a pronounced line. Now that they have a virtual monopoly, newspapers realize that their position would not be well tolerated if they espoused a partisan point of view. The organs of the press therefore make their columns available to several currents of opinion during an electoral period. They avoid supporting ideological positions too far from the central ground. The press plays only a modest role in supporting diverse regional feelings.

Public opinion avails itself of other channels. It is expressed through associations and the bulletins or pamphlets they distribute. Ecology is very popular at local or regional election time; this is due to groups resisting the implanting of new nuclear sites, calling for the stricter control or closure of existing ones, for organic farming and for nature conservation. People are sensitive to the denunciation of dangers which are not in the least abstract and concern everyone.

In the provinces public opinion has for long been channelled by several well established institutions: the Catholic Church possesses solid advantages. Other opinions, unlike it, do not have representatives implanted everywhere and channels to diffuse information, but they can count on the devotion of political or trade union militants. Regional political life was thus polarized by the confrontation between the dignitaries who gravitated around the Church and those who were supported by the anti-clerical left. It was a question of stirring people up by inciting indignation, disapproval or enthusiasm, to the extent that local politicians pay the greatest attention to popular moral pressures.

In western countries and in certain Third World countries today's churches have lost much of their means of exercising influence. They have given up defending social causes they used to be associated with. They have broken with the conservatism which profited from the moral code that they guaranteed; they willingly support protest movements. In losing their traditional foundations they have liberated the local scene: diverse political opinions appear and debates become animated. The associations and new means of communication have permitted the appearance of a new type of leader of public opinion: the local intellectual, teacher, engineer, bookseller, retired person. This figure inspires regionalist movements, fights for the preservation of the historic heritage or passionately defends the natural environment.

From now on, these currents of opinion, of which this person is the conscience, weigh heavily during election campaigns.

Democracy and the role of influence in local and regional political life

The overwhelming majority of people are very attached to local politics. One has the feeling of direct participation and of wielding significant power. One knows the mayor, the councillor or the member of parliament; one bumps into them in the street, they greet you. They are close and open to discussion. It is not intimidating to meet them, explain one's problems and ask for their intervention.

The proximity of power is indispensable to the vigour of territorial democracy: one immediately hesitates to embark on administrative regroupings that the increase in mobility and the size of the framework of daily life nevertheless necessitate. The majority of the 36,000 French communes are too small to have the means needed for the preparation of planning decisions, for the elaboration of town plans or the protection of the environment. The division into minute administrative areas, on which daily migrations confer an obvious unity, generates major social distortions: certain categories pay only a fraction of the costs of the least well-off levels.

Many of these inconveniences are avoided by defining larger units but the electorate becomes demobilized. In the countries of north and north-west Europe where the geography of local authorities has been made to conform with functional realities, the electoral participation rate has generally fallen. In France it has been preferred to allow traditional communes to persist. For the tasks that exceed their means they find the solution by constituting inter-communal syndicates.

This often lively local democracy, favours however the role of influence between important actors. Discussions and decisions are made between dignitaries, and the danger is to see clientelism falsify the role of democracy. The commune is close to the elector, but already the region is too remote to be very interested in what happens at this level. Whether there are territorial collectivities at a certain level or not, the indispensable dialogue to improve conditions in economic life involves the same actors: chambers of commerce, representatives of central government and dignitaries. In France, agreements were easier to conclude when they were not surrounded by too much publicity, to the extent that regional economic action was never so effective as during the 1950s before the installation of the first official institutions: the chambers of commerce, employers' federations, local elected officials at various levels and the powerful organ constituted by the *Caisse* *des dépôts et consignations** learned, in the context of expansion committees, to work together to create the necessary conditions for the growth of industrial activities in regions previously neglected.

The changes, which have taken place everywhere (except in Britain) over the last thirty years, have led to giving greater influence to the regional scale and to giving it a formal status as a territorial collective. The choking of central services encouraged this. The transformation of the economic setting, which deprives states of the classic means of pressure that they had over enterprises, has led in the same direction: no longer can spatial planning be achieved by imposing constraints on the decision makers. They must be seduced to lead them to locate in areas of insufficient employment. Since major firms scarcely take on additional workers any more, initiatives that come from below are precious: to identify them and to create the climate indispensable to their success, power must be close by, know the ground well and be capable of taking rapid decisions. At the same time, the rise of cultural and ecological preoccupations also implies policies that integrate the whole set of regional data and their interactions.

The boom in regional institutions is perfectly explicable by the modifications in the global economic and cultural climate. Their role has become capital in the domains of the quality of life, nature protection or professional training. It is not, however, certain that their action is always more effective in dealing with private-sector groups than the informal procedures used a generation ago.

8.5 Regional Policies

Since the middle of the 1950s, there has been much discussion of regional policies. The debates on this topic often suffer from confusion: actions led at different levels of the territorial hierarchy and which have different objectives are ranked under the same label.

The policy of spatial planning by central governments and its regional dimensions

Central governments must monitor their nation's security, give it the best chance in the concert of nations, develop its economy, increase its general standard of living and provide all the public services necessary for the flourishing of a contented society. These objectives require a global view of the equipping and developing of space. Nations

* A state savings bank which can give low-interest loans for development purposes.

must capitalize on their resources and construct the infrastructures permitting their exploitation, their transformation and the transfer of what has been produced from them. Production must be stimulated and directed by a coherent system of markets, by the establishment of sufficient stockage capacity and by the creation of networks of research centres.

All activities require a juridical framework, which defines the nature and form of the various types of enterprise, their rights and responsibilities as well as the conditions of employment that must be offered to the workforce. They can only survive if they have adequate financial resources – this supposes a financial market and the efficient working of the credit system to supply the needs of working capital. In a world that has become more competitive and where firms often have plants located far from their headquarters, sometimes in foreign countries, it is necessary that optimum contacts are established, demanding that the national space has at least a capital of international rank and internal communications lines (air or high-speed train) making it accessible from all the dynamic towns and regions in the country.

The national policy for planned development of the territory must determine the general structure of networks and guarantee the articulation of the internal space with the outside world by equipping the centres necessary for the nation's presence on the world scene. It must monitor resources, avoid their wastage, limit pollution, safeguard the heritage which cements national consciousness, and act so that natural species that exist only in their country escape destruction.

National regional planning policies are often based on moral considerations, on the vague idea that it is better to look for harmonies or balances, than on a real understanding of the processes at work in spatial organization. Until the Second World War, rural depopulation so impressed the imagination of politicians and seemed to compromise the fundamental traits of our civilizations that actions were aimed above all at restraining the growth of large towns and favouring a return to the land.

Just after the Second World War, the impact of Colin Clark's ideas let it be understood that the decline in agricultural employment was inevitable. But the idea of a fundamental disequilibrium that must be corrected continued to dominate spatial development policies. What changed was the strategy to be followed to achieve a more equitable distribution of population, wealth and initiatives. To stimulate provincial life, it was decided to disperse industry and create the conditions to locate it in zones previously neglected.

It was noticeable at the beginning of the 1960s that newly implanted industries did not always succeed. It was thought that it was because of the lack of well-equipped service centres that many firms or enterprises leaving Paris failed. In the European countries that have best sustained their growth and industrialization (and one always thinks of Germany), the regional capitals are powerful. To provide for the country's development it is necessary to create the framework of metropolitan cities which is lacking. This doctrine borrows from central place theory, but does not have its significance or boundaries. The idea of the growth pole, launched by François Perroux, has been willingly adopted.

The actions inspired by these doctrines did not give the expected results. The doubt that set in led on to a new approach: it is from 'bottom-up' development that decisive initiatives are now anticipated. From this viewpoint, the second thrust of regional policies assumes a growing prominence: the creation of territorial structures capable of stimulating local initiatives.

Regional policy conceived as the establishment of local or regional frameworks

Many specialists consider that the role of central government is not to intervene directly in the spatial organization of the economy, but to put in place the territorial structures best designed to stimulate indigenous energies and to provide locally the support and linkages they need. Regional policy is ceasing to find its mechanisms in the economy. It intervenes through administrative reform: the hierarchy of territorial divisions must be adapted to the new scale of local life and the increased role of medium-sized towns or metropolises: the local coordination of action must also be improved. France is seeking to achieve this by the management of territorial collectives in place of traditional administration.

Once the new entities are established and conferred with extensive powers, they must be assured of the necessary means to carry out the planned development of their area, improve the quality of life and encourage initiatives and dynamics. The problem is that it is the poorest collectives which have the greatest need of facilities, tax concessions for enterprises that move in, and funds for social programmes. Certainly, they can impose a higher taxation rate than elsewhere: the effort of growth must come in the first place from the local population. But there are limits to the pressure that can be exercised, or else negative effects appear. Local taxes at too high a level drive firms away and initiate a spiral, which in time translates into a reduction in the revenues that can be taxed. If the central state wishes that the regional action that it intends to favour by establishing decentralized territorial structures should produce results, it must make transfers and support the expenditure of the poorest regions by the richest ones.

The action of regional authorities

When considering regional policy, one thinks of actions led by the regional collectivities themselves. In similar cases, the possibilities offered to them are different according to the level of the territorial division. It can happen that all the decisions are taken at the provincial or regional scale, and that below this we find only minute districts with responsibility merely for the provision of basic services.

Situations where responsibilities are carefully distributed between several levels are more frequent: this is the case in France. The region has powers of action only in restricted sectors. Roads, for example, are the responsibility of communes, *départements* and the state. The region is not however totally powerless in this area, it has the ability to contribute to the financing of road links that seem a priority for it.

The severely restricted nature of regional powers in the end makes the task of enacting coherent development policies rather difficult. The coordination of initiatives that one has a right to expect from decentralization is often impossible, because it demands agreement from too many partners. There are, however, domains where the regional authorities enjoy by no means negligible funds, and do not come up against institutional competition. Examples are, funding for research, for certain aspects of cultural policy and for environmental protection.

The institutional framework thus imposes a certain style on the policies led by regions. They intervene less by localized actions than by sectoral aid, by taking initiatives, by the effort of training or by encouraging certain types of research or development.

In the present economic climate, the major towns play an essential role in growth: labour markets must be very extensive if they wish to attract cutting-edge activities and back sophisticated technologies. Rapid transport takes place by air or high-speed train: in either case, these facilities are only possible if the population served is sufficiently large. It is thus by action in the metropolitan areas that the biggest effort is made to transform the conditions of regional economic life.

The regions possess only limited power in these domains. It is difficult for them to allocate too great a proportion of their resources to the largest towns: the representatives of the rest of the regions watch jealously to make sure that funding is distributed in an egalitarian fashion. The major agglomerations nevertheless are the essential instruments of regional policy. The state is conscious of this and accepts responsibility for certain of their facilities, technopoles,* airports,

* Technopole refers to science and business parks based on state-of-the-art technology, for example electronics or bio-technology. A link with university or government research laboratories is often involved.

téléports,* public transport networks, etc. It exceeds the regional dispositions that it created. The policy of planned regional development transforms into national intervention favouring the city.

Conclusion

The interest in the region was born of the desire to build administrative divisions on a scientific basis to remedy the anarchy that often prevailed in this domain under the monarchies of the *ancien régime*. By a curious throwback, geographers have professed to ignore the official regions born of these efforts, because they seem to them to be poorly calibrated with the spatial interdependencies which result from their analyses. They also perceive the contradiction between the rigidity of the administrative outlines and the functional structures that they have demonstrated: these were constructed around poles or foci that were easy to define, but their limits were essentially fuzzy.

Since the end of the Second World War, the bad feelings between geographers and the administration have ended. To plan the national space better and to secure prosperity and increase the economic dynamism of countries, to make the centres of political decision closer to the territories the organization of which depends on them, has seemed the most satisfactory option.

In this respect, it is not sufficient to know how to define boundaries and hierarchies by taking into account real-world interactions so that political initiatives might help the whole of the territories to flourish. What the regional level can contribute to the harmonious development of the country depends on the choice between centralized management and management by territorial units, and the complex relationships which bond the various levels of the territorial hierarchy. The general tendency is to bring power closer to those subject to it. The operation contains both advantages (a less oppressive bureaucracy, greater flexibility and an increased ability to create conditions for a dynamic economic and cultural life) and disadvantages (the dangers of clientelism and the exaggerated strength of local leaders, among others).

Further Reading

Agnew, J., *Place and Politics: The Geographical Mediation of State and Society*, London: Allen and Unwin, 1987, 267 pp.

—— and Corbridge, S., *Mastering Space: Hegemony, Territory and International Political Economy*, London: Routledge, 1995, 260 pp.

* Téléport refers to advanced communications facilities, such as satellite links and fibre optics.

Barth, F. (ed.), *Ethnic Groups and Boundaries: The Social Organization of Cultural Differences*, London: Allen and Unwin, 1969, 153 pp.
Eisenstaat, S. N., *The Political Systems of Empires*, New York, 1963, 524 pp.
Fortes, M. and Evans-Pritchard, E. E. (eds), *African Political Systems*, London: Oxford University Press, 1949, 302 pp.
Middleton, J. and Tait, D., *Tribes Without Rulers*, London: Routledge, 1958, 234 pp.
Muir, R. and Paddison, R., *Politics, Geography and Behaviour*, London: Methuen, 1981, 230 pp.
Paddison, R., *The Fragmented State: The Political Geography of Power*, Oxford: Blackwell, 1983, 315 pp.
Painter, J., *Politics, Geography and 'Political Geography': A Critical Perspective*, London: Arnold, 1995, 206 pp.
Planhol, X. de and Claval, P., *An Historical Geography of France*, Cambridge: Cambridge University Press, 1994, pp. 119–468.
Taylor, P., *Political Geography: World Economy, Nation-state and Locality*, London: Longman, 3rd edn, 1993, 360 pp.
Wannop, U., *The Regional Imperative: Regional Planning and Governance in Britain, Europe and the United States*, London: Jessica Kingsley, 1995, 450 pp.

PART III

THE FORMS AND EVOLUTION OF REGIONAL ORGANIZATION

9

The Evolution of Forms of Regional Organization: Societies Without a State

The relations that societies form with the environment and the way in which they develop and shape the territories they live in depends on the ensemble of production and transport techniques at their disposal as well as the systems of social organization which enable them to master, more or less effectively, wide expanses.

In certain regions of the globe, the population still appeared, at the start of the nineteenth century, like a loose mosaic of badly structured small groups. Their cultures were characterized by two types of limitations:

1. production had recourse to methods with poor returns, to the extent that densities generally remained low, and agricultural surpluses, without which it is impossible to sustain the dominant classes, were rare;
2. the institutions were unable to accommodate within the same social structures large populations controlling wide expanses.

When ethnologists painstakingly study their lives, these disdained savages reveal their richness and the diversity of their cultures. One is astonished by the intimate knowledge of the environment which permits these groups, lacking sophisticated tools, to exploit meagre and obdurate resources. The kinship systems, the myths and religious beliefs, the oral literature or certain art forms testify to the ingenuity and fertility of groups which did not know how, or were unable, to develop effective forms of exploitation, and knew how to settle internal conflicts by palavers and masques, thus escaping the cumbersome order of large political structures. Stateless societies, or those with very imperfect forms of statehood, dominated the vast majority of America at the time of its discovery; they were ubiquitous in the forestlands of Africa. The more structured groups which eliminated them from the savannas did not generally possess much more effective material techniques and were unable to create stable and strong states. The Oceanian world belongs to the same group. It owes to its very ancient traditional agriculture, densities which are often high but, at least in Melanesia, have never evolved towards state institutions. These societies have changed much by contact with the West, but they retain enough of their specific characteristics to stand out.

9.1 A Limited Diversification of Landscapes

The landscapes created by archaic societies are generally relatively uniform and their humanization remains restricted. This is related to the techniques employed. Many groups are content with hunting, fishing and gathering. In other cases, in pre-Columban North America for example, crops supplied only a variable, and often modest, proportion of foodstuffs. The essential impact of pastoral peoples on the environment results from the periodic burning of grasslands – the young growth is more suitable for livestock.

Agriculture, where it dominates, depends on manual labour and does not use draught animals. This limits the area exploited by each household (they are usually less than one hectare and are often described as gardens rather than fields, as in Melanesia). The absence of manure makes soil restitution rather difficult. To restore fertility there is no alternative but to practise long fallow periods. To prepare the land for cultivation the vegetation is fired. This is done without preparation in the savannas, but in the forest zone requires that young trees be felled and large trees ring-barked. Thus the fields have a bristly appearance, dotted as they are with trees that are dead but still standing. Since draught animals are not used, there is no reason to cultivate in furrows. The geometric patterns that we are used to are often absent. Generally several species are planted or sown on the same plots: attacks by erosion are thus inhibited. Plants that are intolerant of sunlight find shade. This accentuates the impression of disorder, which prevails on these lands but does not mean that they are neglected.

Basic groups and landscapes in Melanesia

This is how Malinowski[1] described the landscape of the Trobriand Islands, specifically the principal island (figure 9.1) in 1915, before external contacts had altered it too profoundly:

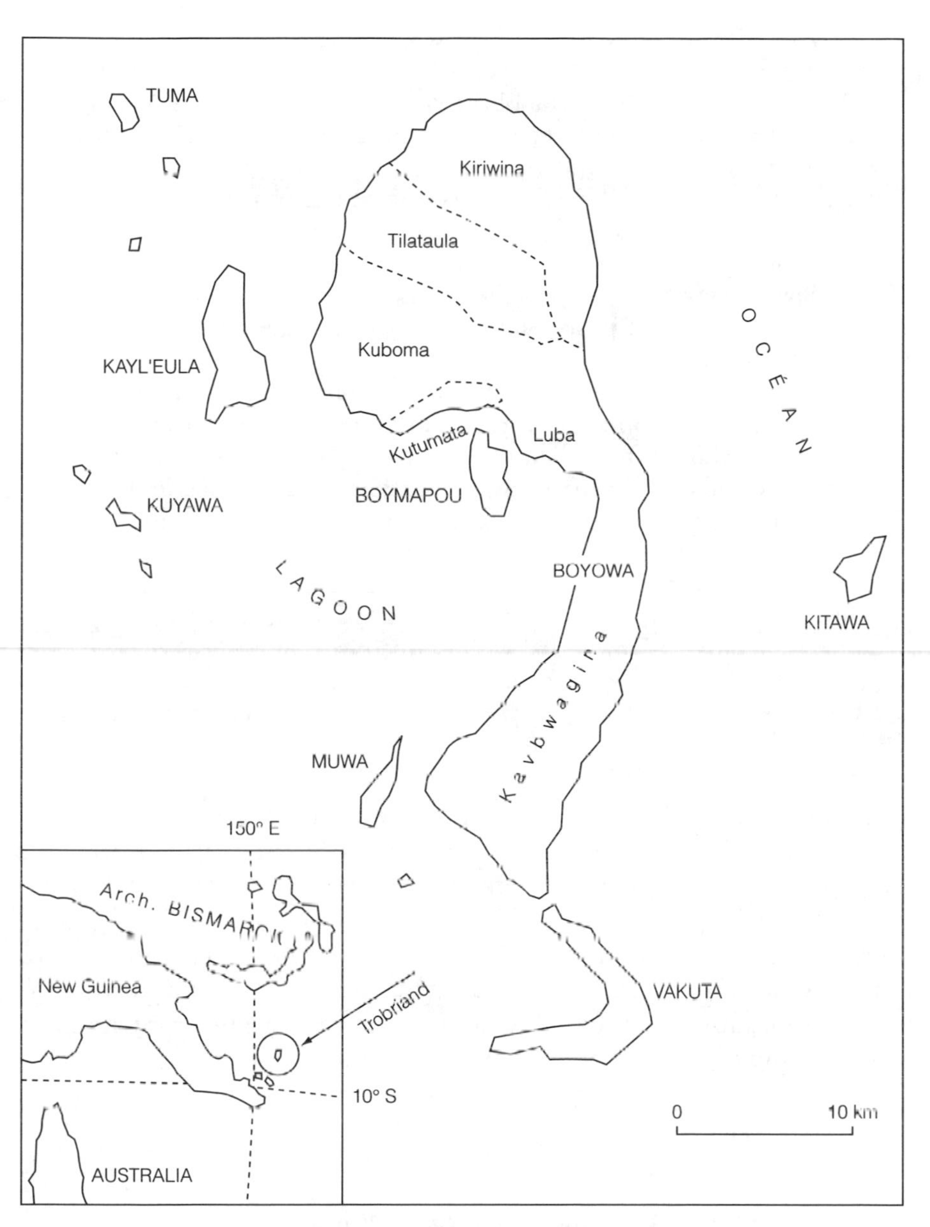

Figure 9.1 The Trobriand Islands
Boyowa, the principal island of the Trobriand is situated between the sea to the east and the lagoon to the west. A regional differentiation is discernible in the northern part.
Source: B. Malinowski, *Les Jardines de Corail*, Paris: Maspéro, 1974, p. 8.

At first sight, the landscape is not beautiful. . . . The . . . land is almost entirely devoted to shifting cultivation, so that the bush, cleared regularly at intervals of several years, never achieves substantial development. When one walks in the island, sometimes one passes between

> two walls of dense, low jungle, regrown only recently, sometimes one crosses gardens.
>
> The gardens are without doubt the beauty of the landscape. We cross a completely cleared space, open to the horizon, where here and there we discern a copse identifying a *boma* (sacred wood) or one of the island's innumerable villages. . . . This is what one observes at the onset of growth: high tufts of sugar cane, young, stringy leaves of *taro*, here and there, an early shoot of *kuwi*, a large variety of yam, its tendrils climbing stalks which escaped the clearing and burning. In the foreground, we notice the posts which already divide the gardens in a vast draughtboard. . . .
>
> Now we cross a garden of fully grown yams: it bears a remarkable resemblance to a hop field in Kent. Around the massive stakes exuberant tendrils climb whose dense garlands of leaves rise and fall like fountains of greenery, producing a sensation of abundance and shade often evoked in the magic rituals of the indigenous population. Even the gardens, where harvesting has already taken place and where in places a banana tree and old sweet potato plants remain, have the charm of a badly maintained orchard. In marshy areas, we may pass alongside a *taro* garden, dotted with scarecrows and rattles, whose carpet of foliage is surrounded by a solid new fence. . . . At a rough guess, I estimated that a fifth or perhaps a quarter of the total area was simultaneously cultivated. (Malinowski, 1974, pp. 61–2)

As we see, the cultivation is conducted with much care. The number of varieties, adapted to the slightest differences in the environment, and the diversity of plants on which the lives of the group are based, indicate long and rich traditions. In general in archaic societies, however, the most important modifications of the landscape are due to repeated burning, to the place held by secondary vegetation forms and to the propagation of species by cultivation – or dispersed by animals when rearing is also practised. Once cultivation is no longer practised, the fences which protected the fields from wandering animals are flimsy and quickly disappear.

Basic groups and landscapes in forested Africa

In Gabon (figure 9.2), Roland Pourtier[2] notes: 'It is scarcely a paradox to say that there is not a "landscape" in the underpopulated forest environment. The instability of localities, the evanescence of the traces of mankind makes a landscape approach inapplicable and cartography hazardous' (p. 147). The land is practically empty, with tiny units of population, a hundred inhabitants on average, grouped in small villages. The grip on the environment is multiple, but difficult:

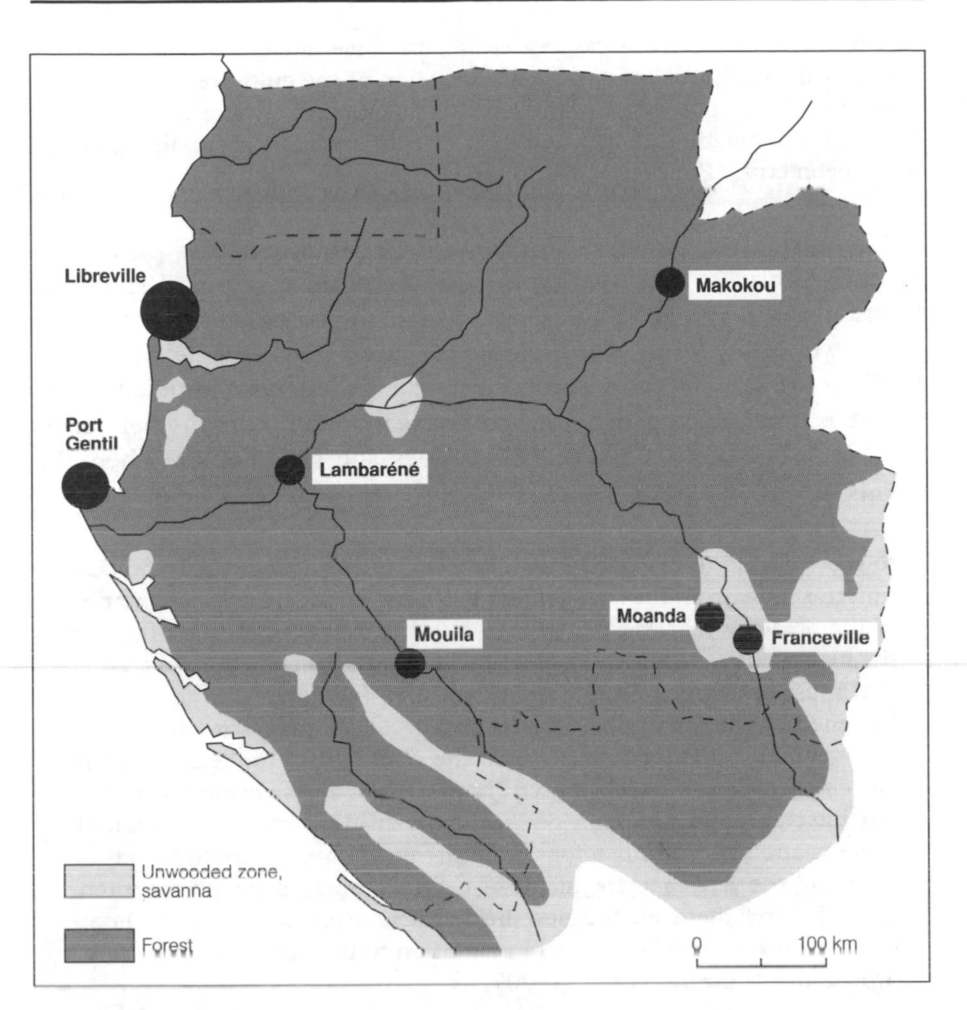

Figure 9.2 Gabon
Source: R. Pourtier, *Le Gabon*, Paris: L'Harmattan, 1989.

> Scarcely have [the limits of the village] been crossed, than one penetrates into a less and less dignified ensemble as one moves further away. In the immediate proximity of the houses, a narrow belt of permanent crops with bananas predominating, surrounds the village over a depth of several metres. Afterwards comes the space devoted to agricultural exploitation, which because of the dominant function we may term 'the lands'. Its appearance in no way evokes the ordered lands of a countryside, but much rather a confused brush. The fields, communally called 'plantations', alternate with fallow lands varying in aspect according to their age, and remains of a more or less old forest. Crops and recent fallow are easily identified; on the other hand, the transitions between old fallow and regenerated forest escape the inexperienced eye. With time, the recrudescent forest effaces all traces of clearings. The

> dispersion of burned areas ceaselessly modifies an essentially mobile landscape composed of a cyclical succession of the cultivated plant and the tree. The appearance of the lands is like a puzzle where the moving trilogy of plantations, fallow and forest are imbricated without the slightest geometry. (Pourtier, 1989, p. 162)

The agriculture practised on the village lands employs only rudimentary tools: 'In fact, two types of tool are necessary and sufficient to practise a forest agriculture: one to cut the trees, a felling axe, the other to clear the scrub, weed, dig up the crops, the machete' (Pourtier, 1989, p. 182). The man is the clearer: he prepares the felled wood which is fired during the brief dry season. The woman then takes over: it is she who cultivates. How does she set about it when she decides

> to grow a 'field of manioc'? First of all she takes cuttings, 40 to 50 centimetres long, from the stems of adult plants, that is from the plantation of the previous year, she loads them into her basket, transports them to the newly-burned field which often still has incompletely burned stumps that must be stepped over. With a rapid blow she opens up the ground loosened by the fire to plant her cuttings. . . . The plants are distributed non-geometrically but in relation to the obstacles of the terrain and in any case sufficiently widely spaced (between two and four metres) to permit inter-cropping. Another day, the woman digs from her plantation of taros and yams destined for the family meal, taking care to retain the collars of the yams and fragments of taros carrying buds with the intention of using them on the new field. Sugar cane and banana shoots will be transplanted from the old plantation to the new one by the same procedure. (Pourtier, 1989, p. 207)

The disordered and uncompleted nature of these tropical croplands is easily explained. The fields bear a single crop: at the end of one, two or three years they return to fallow. On average, twenty ares* of crops are sufficient to feed one person; with fallows of the order of twenty years, it takes two hectares per inhabitant, taking into account all the stages of the recovery of the forest cover. Since not all the forest is capable of being developed, four hectares is a more realistic average.

This zone where the impact of groups is detectable, if difficult to define, is surrounded by a larger aureola which 'corresponds to a hunting space' (Pourtier, 1989, p. 162). The forest is a difficult environment to clear:

* In the metric system an *are* equals 100 square metres. It is specific to the measurement of agricultural land areas.

> The understanding of the components of the forest and their location is a *sine qua non* condition of appropriating it. Years of learning are necessary before the initial chaos resolves into a meaningful order, so that the forest, which is no longer an informal aggregate of trees, reaches a level based on knowledge. . . . The ability to interpret the forest is fixed by the limits of familiar space. . . . The forest is uncovered better and better metre by metre, without gaps in the continuity of the cover. In an enclosed environment which severely restricts the view, progress forward implies that landmarks must be established behind. The extent is thus narrowly circumscribed by the need to learn and memorize the outlines of each of its places. (Pourtier, 1989, pp. 149–51)

The areas exploited for hunting are thus limited and roughly circular in structure around the village.

This is how the human perimeter appears in the forests of Gabon. As Roland Pourtier explains:

> this schema implies a low density, the absence of lines abutting onto two adjacent land holdings. This is the general case. Each inhabited cell possesses an expanse sufficient to extend a protoplasmic territory. The area exploited from each village reproduces the same concentric model displaying a negative gradient of adaptation away from its core. Bathing in a forest backcloth, the human groups are arranged according to their internal dynamic, indifferent to the uncertainty of their confines. (Pourtier, 1989, p. 162)

Tasks demanding high density

Human impact on the landscape becomes never so obvious as in the case of exceptions. In the Philippines, there are tribes, like the Igorots, who practise irrigated rice-growing on terraced fields. Entire hillsides are equipped in this way, often right to the summit, with no natural vegetation left.

In West Africa, there are cases where areas of different relief, rocky and discontinuous soil cover, are the object of carefully tended agriculture, with terraces, with repeated working, a high labour input and an intensification of land use leading to a shortening, even elimination of fallow. These enclosures in the middle of the extensive landscapes of the rural Sudanese civilizations betray the presence of palaeonigritic groups: refusing the techniques of their neighbours, keeping faith with segmented structures, they can only survive in sectors disdained by the dominant groups, and are obliged by their density to intensify their crops. The best-known example is that of the Dogons of the Bandiagara scarp, in Mali, but equally we may cite the Kabayé of Togo.

The zones where landscapes are profoundly modified correspond with situations of contact with more advanced culture groups. The original settlers could not resist pressure from their neighbours except by adopting techniques which allowed them to increase in numbers – this is the case of irrigated rice cultivation in the interior mountains of Luzon in the Philippines – or by adopting novel methods of intensification, as can be seen amongst the palaeonigritic peoples. When they are not submitted to outside pressure, the peoples who oppose organization based on states generally avoid high densities: they impose too many contacts, encourage the emergence of conflict situations and multiply the opportunities where an intermediary might impose his power.

Trade and the organization of space: the African forest lands

Trade takes place between settled groups but specialization remains limited. Transport is too difficult to haul the grains or roots that are the basic foodstuffs for more than a few kilometres. Salt, dried fish or meat travel better, but are never the essential basis of food supplies. Materials necessary for making tools, or tools themselves, are acquired externally. Also involved are prestige goods: the low weight and the high value invested in rare shells, striking feathers, stones or precious metals, make them highly mobile.

In some cases, self-sufficiency is almost absolute: this is the case in the African forests analysed by Roland Pourtier in Gabon. The roots and plantain bananas which constitute the basis of the vegetable diet cannot be transported far; this is even more the case for meat obtained by hunting. 'Without barns, without markets, the people of the Gabonese forest have built over time and over space the characteristic relationships of "self-sufficiency" (Pourtier, 1989, p. 212).' The basic unit organized around each married man provides practically all its needs and can survive alone. The combination of several units permits cooperation, and due to solidarity can survive difficult periods. The village, although small, is nevertheless above the scale where self-sufficiency is achieved.

In the case of Gabon, relations between neighbouring cells are motivated much more by social reasons – for the exchange of women – then for trading contacts. For a long time, iron was the only product for which the village could not supply its needs. Western arms and tools were added later, but without stimulating very active trading currents.

The Gabonese space is thus structured at several levels: that of the villages and their lands first of all. The densities are so low that there is no superimposition: almost everywhere substantial tracts of unexploited forest remain between the inhabited homesteads, which from the point of view of food supplies constitute self-sufficient units. These buffer no-man's-lands are crossed by a few trails: contacts are forged at a matrimonial level, expressed by frequent visits and gifts: 'Matrimonial zones on the contrary are imbricated on each other from a number of centres and cannot be used as a criterion for the division of space' (Pourtier, 1989, p. 245).

The significance of ethnicity remains to be considered:

> before the formation of the nation-state, ethnic groups form the most widespread level of identification. But it would be erroneous to analyse them by reference to the national model: ethnic models are not 'micro-nations'. The sub groups defined in terms of cultural or linguistic origins, situate individuals but are not collectively projected. These are reference systems, not systems of frameworks. One thing is in fact remarkable: there has never been, as far as we know, ethnic power, institutions or chiefs representative of an ethnic group. (Pourtier, 1989, p. 247)

The forest universe is thus one of fragmentation, of tiny cells, linked by a skein of footpaths without a dominant axis, an impression of a blurred world, that is the feeling one discerns. Only one level is really important in people's lives:

> A privileged level exists through the life of social groups. It has no name and is difficult to define. It does not coincide either with territory or ethnicity. This space is that where active solidarities operate, that of connivance between places and beings, something like a *pays* which takes its consistency from the effect of repeated journeys and sustained alliances: a familiar *pays*, peopled with signs, traversed by memorized journeys, domesticated and whose inhabitants are identified and identifiable. In consequence it is a space of small extent, at the scale of abilities and practices, that the forest environment tends to constrain, at the scale which makes solidarities so effective, a space of nearness. But no *pays* is enclosed within precise and stable limits: fluidity remains the rule. (Pourtier, 1989, p. 247)

Reflecting on the traditional region in Black Africa, G. Sautter[3] notes:

> Anthropological research has revealed the existence, and often the force, of political structures founded either on extended networks of kinship, or on varied associating groups, or even places of religious character. These structures have the effect of solidifying, within the geographical framework that they fit into, both men and the different parts of space

> they occupy. But the mesh that results, although in many cases contributing towards distinguishing a region as compared with amorphous space ... which the environment does not structure in the sense of a hierarchy and complementarity corresponding to the fully developed model of the region. (Sautter, 1968, pp. 82–3)

African societies without states are not unorganized, but the contacts which unite people there are so different from those that we are familiar with that it is by their levels and forms that the solidarities which structure them may be interpreted.

Contacts and spatial organization: Oceania

In the Trobriand Archipelago, the difficulty of trade contacts prevents developed forms of specialization. Everywhere the essential foodstuffs must be produced locally. Differentiation therefore above all flows from the adaptation of ways of life to the environment. Malinowski expresses this very well for the principal island of Trobriand:

> in some villages of the lagoon, fish is the principal food source of the inhabitants, and this activity absorbs half of their time and effort. But if fishing is important in a few regions, agriculture reigns in all of them. If a natural or cultural catastrophe should prevent the Trobriand people from practising fishing, agriculture would suffice to feed all the population; but when drought strikes the gardens, famine is inevitable. (Malinowski, 1974, p. 15)

A certain degree of specialization is however discernible, and partly explains the division into regions that Malinowski recognized on the island:

> Thus, in the North, there are three central provinces [figure 9.1]: Kiriwana in the north-east, Tilataula in the centre, Kuboma in the south-west. The two first live exclusively from agriculture. Kiriwina is the dominant province in political terms, the most elevated socially and perhaps the richest economically. ... In Kiriwana, the majority of the villages have a maritime 'base' on the eastern coast where a sea-going dug-out canoe is grounded on the beach and embarkations are made for fishing and for cabotage.
>
> The next province, Tilataula, does not practise fishing at all. ... Its inhabitants are proud of their agriculture and their frequent victories over their neighbours. ... But it is only when we visit the Kuboma district, further west, that we encounter truly developed industries on the rocky soil where the gardens are far from having the splendour that their neighbours to the east have. (Malinowski, 1974, pp. 19–21)

Economic potential exists then, but on a common basis which must not be overlooked and which contains specialization within very narrow limits. Landscape differentiation is equally retarded by the absence of real hierarchical structuring. Contacts established between villages are numerous, but follow various directions. As there are no political superstructures, there are no tributes or taxes raised to profit a coordinating centre on which all the routes converge. And material exchanges span short distances, permanent service centres scarcely exist to aid their organization. Important meetings sometimes take place, prestige items (those which motivate the *circle* of the Kula, that Malinowski cites for example) are exchanged on those occasions but nothing reveals externally the role of these places once everyone has departed.

9.2 The Psychological and Cultural Dimensions of Territorial Constructions

The apparent uniformity of landscapes should not lead to the conclusion that there is an absence of large-size territorial constructions. It is true that life takes place essentially within the basic groups constituted by the local communities. One belongs to a village when the group lives from agriculture, or to a small group, in the case of a pastoral tribe. But the realities of a larger dimension exist in people's minds. They are expressed where the local differentiation of traditions facilitated by oral transmission is strong, by the existence of areas speaking the same language or the same dialect. The cement which binds unity is often ethnic: people feel close together because they are descended from the same ancestors, believe in the same supernatural forces, believe in the same fables. But the binding is also geographical. The identity of many groups is founded on the land: the collectivity exists because its life is set in an environment on which it has left its mark profoundly and which communicates its distinctiveness to it. Territory building is intimately linked to the ontological structuring of space.

In Vanuatu, territorial consciousness reposes on the idea that one was present at the origin of the world. Joël Bonnemaison[4] writes:

> The sacred world corresponds to that of the founding heroes who, in . . . the 'time of dreams', shaped the universe, established the social laws and distributed in certain places magical and supernatural forces. The landscape is still the invisible tracing of this fantastic epoch where everything assumed its meaning and the universe its definitive forms [figure 9.3]. . . . The passing from chaos to order is thus at the heart of the majority of the great mythologies which circulated in the archipelago. As a general rule, the legends explained that in the beginning

> a cosmic substance pre-existed, shapeless and soft which 'supernatural beings', part men, part beasts, part spirits came to shape and transformed it to its present appearance. They made various species of plants appear, created night which interrupted the monotony of time, the mountains, the rivers, the sun and moon, islands and headlands. They initiated 'customs', celebrating sacred rituals. . . . When these supernatural beings had finished the creation, fatigue overwhelmed them; they departed from the Earth's surface or were metamorphosed into mountains, stones, rocks: the magical times thus closed. (Bonnemaison, 1986, pp. 183–4)

In such a system, social ties are founded on belonging geographically more than on lineage: what brings people together is the shared mythical past.

> In a type of society where the symbolic dimension is stronger than economic concern, expanse effectively counts for less than the anchoring effect. Places of significance form a spatial–symbolic frame that land is organized around; a protective zone of forest extends beyond that which in the past served as a limit. (Bonnemaison, 1986, p. 190)

The principles which are valued in Vanuatu are replicated throughout the Melanesian area:

> Melanesian society (and beyond that apparently all Oceanic society) is as much, and perhaps more, a society of co-residents, united by their links with places, than a society of relatives linked by ties of blood. This 'patriotism of the land', given to the earth which nourishes them, carrier of magical properties and supernatural powers, leads in fact to the constitution of a society on a 'geographical' base. (Bonnemaison, 1986, p. 190)

In Tanna, in the southern part of the archipelago, the myths linking men to the land connect twice. At the origin, in the 'time of dreams', Wutingin, the spirit who created the Earth, sent to the island, to consolidate it, a horde of screaming stones, the *kapiel*, which made great journeys. Stopped in their flight, some of them punctuate the sacred space today. After the arrival of people, the crucial event was the murder of a malevolent hero, *Semo-Semo*, whose corpse was dissected: each portion, bearing a name designating both a territory and the people who settled it.

The people of the island thus share a common mythology, but are organized in smaller territorial units. In an insular world like this, metaphors drawn from navigation are naturally expressed: groups and their territories are named after dug-out canoes. Each one normally consists of an ensemble which starts from the shore and extends as far around the central watershed of the island from north to south. Each canoe groups smaller units. Each one is structured around a holy place, the site of dancing, the members of the entire canoe group recognize each other at the dance site where all the sub-groups can assemble.

It is obvious that groups have ecological and economic foundations since they establish a solidarity between the coastal people and those of the interior, whose economies are to a certain extent complementary:[5] the bushmen of the mountainous interior cultivate *taro* and sweet potatoes, the only root crops that prosper in an already cool environment, practise gathering and much hunting. As a prestige item, they possess pigs. The *sol wara* of the littoral bring the more productive yam, have coconuts, build dug-outs for fishing and can supply turtles, another item coveted by everyone (Bonnemaison, 1990–1, p. 121).

The territories do not live an introverted life. Between neighbouring groups on the same island, it can happen that there is a permanent state of war, but alliances are forged with those that are most willingly visited. 'One of my first surprises in Melanesia,' comments Joël Bonnemaison, 'was to realise that the groups of the east coast of the island of Aboa (Vanuatu) were in permanent communication with those of the west coast on the neighbouring island of Pentecost (figure 9.3) whereas they had no contact with the groups on the west coast of their island even though they were not separated by any landward no man's land' (Bonnemaison, 1990–1, p. 123). In a land where distance by sea is not frightening, extremely close links are often established between opposite shores. They arise from trade: they are structured by networks, but are not centralized.

Malinowski[6] has admirably described the *cercle* of the *Kula*, which embraces the whole of the Entrecastreaux Islands and the Louisiade Archipelago to the east of New Guinea. It encompasses the eastern extremity of New Guinea, the Amphletts in the southern part of the Entrecastreau Archipelago, the Trobriand Islands in the north, and Misima, in the Louisiade group (figure 9.4): this extends over 250 kilometres from north to south and 300 from east to west. The prestige goods which are traded are of two types: the *mawali* circuit transports, in an anti-clockwise direction, shell armbands; the complementary *soulava* circuit, transmits in a clockwise direction necklaces of red shell discs. These are produced at the extreme south-east of the circuit, in Rossel Island, which sends them to Misima and Panayat islands, where they are taken over by *Kula*. Armbands are produced in two places, at Woodlark Island and in the west of the main island of the Trobriand.

Malinowski concludes his study of the *Kula* as follows:

> Its novelty resides in the very extent of the institution, from both the sociological and geographical point of view. A vast organization of

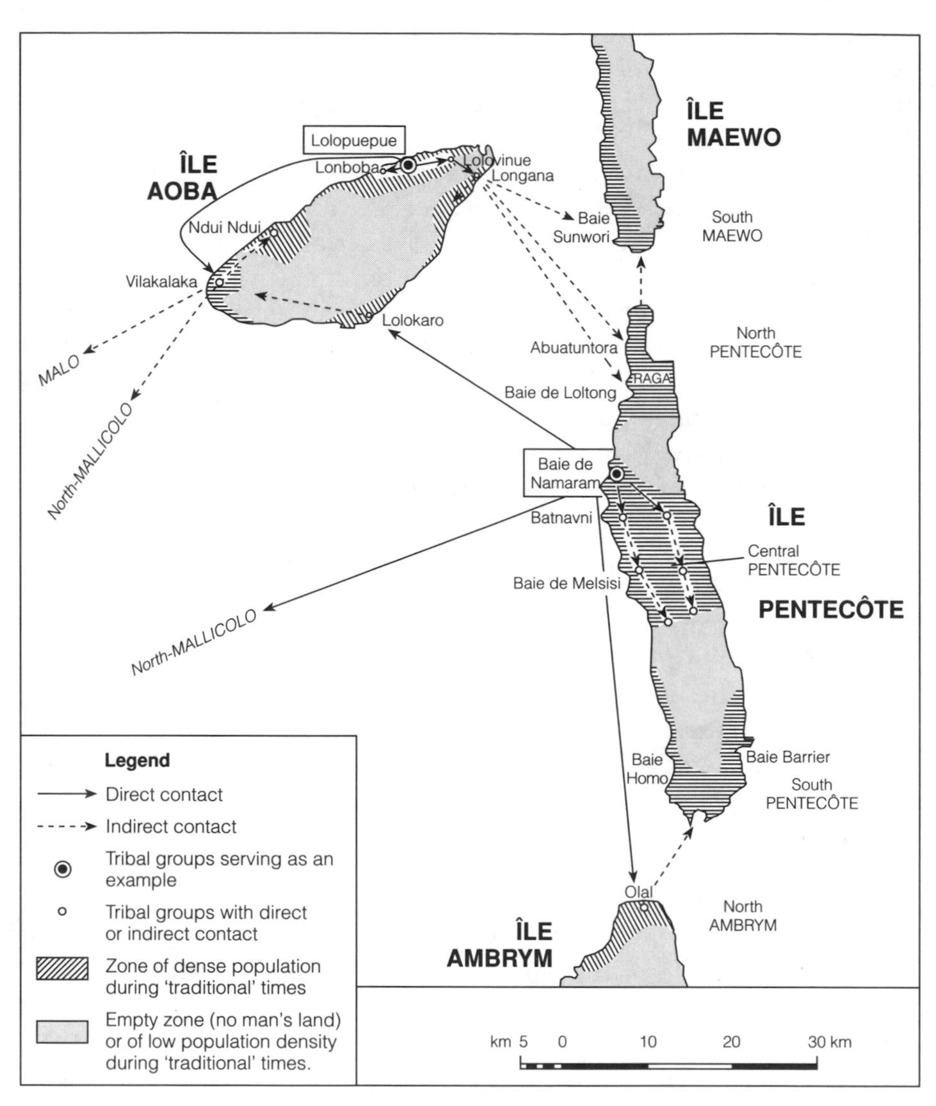

Figure 9.3 The divided space of the Melanesians of Vanuatu: the Aboa and Pentecôte islands in Vanuatu

There are fewer contacts between the two slopes of Aboa island than between the east coast of Aboa and the west coast of Pentecôte island.

Source: J. Bonnemaison, 'Les voyages de l'enracinement', *L'Espace Géographique*, 8, 4, 1979, p. 310.

inter-tribal connections, which, within an immense perimeter, a large number of individuals are bonded by clearly defined obligations, minutely regulated according to a concerted plan, the *Kula* represents a sociological system of a complexity and size without precedent, taking into account the cultural level of the tribes practising it. Furthermore, this important maze of social relations and cultural influences cannot be

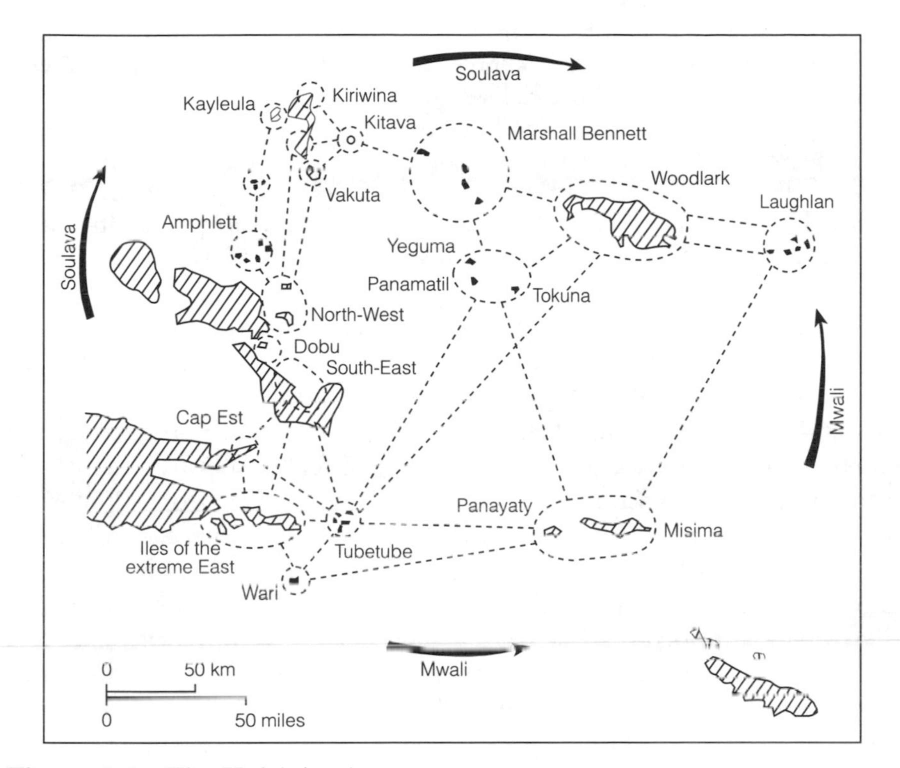

Figure 9.4 The Kula circuit
Source: B. Malinowski, *Les Argonautes du Pacifique occidental*, Paris: Gallimard, 1963, p. 140.

> regarded for a single moment as something ephemeral, recent and precarious. In effect its very elaborate mythology and its magical rites prove sufficiently that the *Kula* is deeply anchored in the tradition of these aborigines and undoubtedly goes back a very long way in time. (Malinowski, 1963, p. 581)

Joël Bonnemaison, who also carries out research in a Melanesian setting, observes:

> in their eyes [those of the inhabitants of the South Pacific], the universe is a myriad of islands in contact with each other. This exploded world in which the island is the norm has no centre: it is composed of networks which project over space a huge net of contacts which bind together in certain places and come apart in others but is forgotten by no one. This netted or reticular space has no centre, it creates a supple fabric whose structure is the mesh. For the islanders, the sea is not an enclosure, but a routeway which creates archipelago effects. Thus no island is really isolated, each is the interface of another.
>
> . . . The people of Pacific islands represent their space as dug-out canoe routes, the shores are parts and the sea an invitation to travel.

> The powers and wealth are in constant circulation, as in the ritual of the *Kula* described by Malinowski in his famous research on the Trobriand (1922): the prestigious shell necklaces and armbands are exchanged for each other and pass successively to an ally, from one island to another forming a circular itinerary. In such a society, the flows do not converge towards a centre whose goods accumulate, but circulate from place to place along a network of alliances, creating a reticulated structure of space. (Bonnemaison, 1990, p. 123)

With access only to simple techniques and lacking efficient means of transport, archaic societies seem condemned to live in isolation. The spatial organization that they fashion testifies, by the place held universally by food production, that the limits on mobility are quickly reached. The absence of overarching political structures translates into and accentuates this fragmentation. It would be wrong, however, not to pay attention to the organizational forms that these people have sometimes been capable of weaving together over great distances: that goods of only symbolic significance are involved does not mean that they are less important. The trade which develops in this way explains the deep cultural similarity that the language and the real difficulties of movement tend to isolate. Even if these great circuits only impinge on a few isolated points in space, they elucidate the common characteristics of large areas of humanized landscape.

The regional geography of zones with ancient cultures, or those scarcely emerging from that state, astonishes the European observer: he is not used to such discreet human imprint, and such a different mix of diversity and homogeneity from that observed in developed countries. The ties which bind together over long distances and give coherence to space where human beings appear to be so frail, have not ceased to fascinate us.

The Oceanic archipelagos owe to their isolation and to the late date of their contact with westerners, the retention almost intact of the upper levels of their systems of relations. It is for this reason that we have chosen to illustrate this section by Melanesian examples: in America it is more difficult to reconstruct the routes and conditions of trade. The archaic groups that anthropologists have observed there were in decline, disturbed by the contacts they had with the great pre-Columban civilizations and by the eruption of colonizing waves.

Further Reading

Allan, W., *The African Husbandman*, London: Oliver and Boyd, 1965, 505 pp.

Barth, F. (ed.), *Ethnic Groups and Boundaries: The Social Organization of Culture Difference*, London: Allen and Unwin, 1969, 153 pp.

Brookfield, H. C. and Hart, D., *Melanesia: A Geographical Interpretation of an Island World*, London: Methuen, 1971, 464 pp.
Chapman, M., Island autobiographies of movement: alternative ways of knowing?, in Claval, P. and Singaravelou, *Ethnogeographies*, Paris: L'Harmattan, 1995, pp. 247–59.
Dalton, G., *Economic Anthropology and Development: Essays on Tribal and Peasant Economics*, New York: Basic Books, 386 pp.
Daryll Forde, C., *Habitat, Economy and Society: A Geographical Introduction to Ethnology*, London: Methuen, 1934, 500 pp.
Malinowski, B., *Coral Gardens and their Magic*, Bloomington: University of Indiana Press, 1965, 356 pp.
Sahlins, M. D., *Tribesmen*, Englewood Cliffs, NJ: Prentice-Hall, 1968, 118 pp.
Service, E., *The Hunters*, Englewood Cliffs, NJ: Prentice-Hall, 1966, 126 pp.

10

The Regional Organization of Traditional Societies

Archaic societies which had no knowledge of writing have left no historical evidence and it is primarily this that distinguishes them from traditional societies. There exists a wealth of information concerning the Egyptians, the Sumerians, the Akkadians, the Persians, the Greeks and the Chinese which gives us a profound insight into their history. Drawing on the work of the turn-of-the-century geographers, however, Fernand Braudel reminds us that the written evidence that we have refers only to the narrow elite representing power, wealth and learning that dominated these societies. Factual history based on diplomatic or literary texts or on works of art which have been handed down to us offers no information on the rural masses who made up at least 80 per cent of the population. In order to reach this bedrock a long-term historical approach is required which takes into account the deep stirrings and rhythms of social groups.

The basis of traditional societies lies, in fact, in the dualism between the elites and the labouring masses, between the town and the countryside, between the often frenzied pace of political events and the evolutions that define the boundaries of the possible and the impossible. Without this dualism it is impossible to understand the methods of organization of space that characterize traditional societies.

10.1 The Transformation and Diversification of Landscapes

Traditional civilizations cultivated the areas that formed the basis of their staple food supply by raising livestock and by agriculture. They did not neglect hunting, gathering and fishing, but these sustained only a very small section of the population and became a sport and a source of income often confined to an elite. Their technological base was undoubtedly superior to that of the archaic groups, but the changes through which the conditions documented by history were achieved often owed as much to the means of communication and support as to those of production. The forging of closer ties between larger groups through the use of force and the influence of the new religions meant that it was now necessary not only to produce a surplus in order to feed warriors, priests and administrators, but for the political system with its specialist staff to become autonomous. It was these surpluses which made it possible for a fraction of the population to become urbanized. The invention of writing facilitated the spread of the new ideologies employed by the political régimes and the keeping of accounts improved the management of the taxes which they raised for their own benefit.

The pressure exerted by the ruling powers led to an intensification of land use, whilst the peace which they provided was conducive to demographic growth. It was for these reasons that the ecological basis of historical societies had of necessity to be agricultural. When pastoral groups succeeded in dominating large areas and creating states they could only function due to a geographically stable and submissive peasantry.

Power brought its own demands and the number of mouths to be fed increased since peace and order were conducive to demographic growth. It therefore became necessary to try to exploit the environment as fully as possible. Across a large part of the ancient world the domestication of animals increased the available work capacity per capita, but this evolution occurred in stages. The introduction of draught animals was a significant factor from early antiquity onwards in the Middle East and in the Mediterranean, but they were only able to make a significant contribution after the invention of yokes and harnesses which rested on the animal's skeletal frame and did not inhibit breathing. It was during the same period in the tenth or eleventh century that the use of metal became widespread. This gave rise to new types of development by providing workers with more efficient tools, but the transition from a hunting, gathering and fishing base to an agricultural base (the so-called Neolithic Revolution of society) was possible without this invention and without the use of animal power. Pre-Columban societies founded powerful states, but their metalworking skills did not impact on their tools and their agricultural techniques relied entirely on human manpower. Their success would not have been so resounding had they not had the good fortune to have in the form of maize the most high-yielding cereal. No other cereal has such a high ratio of seeds sown to actual yield (1 to 50, 80 or 100 for maize as opposed to 1 to 10, 15 or 20 for wheat). Efficiency is equally remarkable when calculated in relation to the surface area, although the gap is smaller when it is measured in relation to the amount of effort. Beans and amaranth were an abundant source of protein in an almost meat-free diet.

The landscapes, characteristic of historical societies, were profoundly transformed by human presence, the involuntary or indirect influence of which is still evident in the effect of fire on forest cover and in the weeds which grow alongside crops. The increase in the population, however, was accompanied by more direct changes. Henceforth the market towns made substantial inroads into the vegetation cover. Their hold, however, was still quite fragile and the emphasis here will be on the new forms of rural landscape.

There were zones in which little or no livestock was raised. It was not always possible to cultivate the land more than once every four or five years on average. This meant that after three years of cultivation the plot of land was left uncultivated and returned to brush of secondary forest for eight, ten or twelve years. Naturally this limited population density, income and town size. A number of the tropical traditional civilizations developed in this way, notably the Mayan empire in the tropical jungle of the Yucatan or Guatemala, the lands of the Deccan in India and certain mountainous regions of South China, Indochina or Indonesia. In Black Africa the zones in which the political structures were the most highly evolved were founded on this type of ecological base, as for example in the Yoruba country in south-western Nigeria.

Elsewhere the influence of agriculture demonstrated greater continuity and was total. Crops took over all of the cultivable areas of land, which were left fallow only once every two or three years. Herds grazed on the stubble after the harvest and roamed the moors and woods which persisted on land that was more difficult to cultivate. In irrigated areas the land was even better utilized. In the Far East rice was grown year after year in the same beds; fertilization was guaranteed by soil restoration and by seaweed that fixed atmospheric nitrogen.

The differentiation of the rural landscape was still restricted by the fact that it was difficult to market staple products over a distance: rural areas had to supply the basic foodstuffs required by the peasants working in them and for the citizens of nearby towns. Diversification took place by adapting lifestyles to the environment rather than taking the gambles for which local conditions seemed to call. Diversification can be seen in land use, dominant crops and in diet, but the ideas underlying peasant life changed very little from area to area. Only where there were very sharply contrasting environments did radically different types of agriculture coexist in close proximity. Where this did occur regional constructions flourished especially well.

Lucien Febvre[1] observed that this was the case in the Franche-Comté in the reign of Philippe II* and he highlights the distinctive features of the region:

> The first and perhaps the most striking is the extreme variation in land in the province and the great diversity of its parts. What do Vôge, Finage, the Saône plain, the Amont or Jura plateaux, and the valleys or crests in the high mountains have in common? Neither their physical appearance, the composition of the land, their climate or produce, less still their inhabitants . . . who by virtue of occupation, customs and character form many distinct groups. . . .
>
> Although apparently so different, these areas were, by the same token, complementary. This juxtaposition within a narrow area of dissimilar districts is a phenomenon particular to France and was especially striking in the Franche-Comté. The two regions of the Jura and the plains were united by a close economic solidarity which was the basis of their political accord. The relationship between them was constant because it was necessary; the mountain regions required grain from the plain and wine from the vineyards and the plains required livestock, wood and sometimes even manpower from the mountains, hence an exchange of foodstuffs and contact between the inhabitants. . . . (Febvre, 1970, p. 29)

On the plateaux and plains of the temperate world, lifestyles were homogenous. Crop rotations and livestock farming were adapted to local conditions, giving rise to unique agrarian and settlement structures. The homogenous zones were often very large; in France hedged farmland covered the whole of the west and open fields the whole of the north and east. It was only where the two areas came into contact that the picture became more complex, revealing secondary contrasts which gave the landscapes individuality.

Rural studies in Europe up until the Second World War often focused on regions not completely transformed by the Industrial Revolution and the transport revolution, where the old economy, which was still discernible in certain features of the landscape, and the organization of crops, reflected the technical conditions in traditional societies. They highlighted many differences, but it was often difficult to interpret their structure on a regional scale. The most significant boundaries for rural life in France – namely those between enclosed farmland and open fields or between the small tiles or slate roofs in the north and the Roman roof pantiles of the Midi – divided highly unified regions such as the Auvergne.

* King of France 1180–1223.

The situation was different where the contrasts between neighbouring areas were greater. In the Far East with its hot, humid summers, the alluvial plains given over to irrigated rice-growing contrasted with the hilly regions where dry crops could not sustain such high population densities and with the mountains farmed extensively by slash and burn methods. The latter areas escaped the control of the powerful civilizations in the lowlands and served as a refuge for less advanced ethnic groups. Indonesia and the Philippines were exceptions to the extent that irrigated rice-growing had taken over a number of mountainsides either because it was a feature of stateless societies (as in the case of the Igrotot of Luzon), or because it demonstrated the manpower and organizational strength of the great agrarian civilizations of the Hindu kingdoms of the Middle Ages.

The Mediterranean world offered equally contrasting conditions: hills and low plateaux well suited for growing non-irrigated cereals and trees; mountainsides which offered plentiful summer pasture but which could not provide sufficient stocks of fodder for the winter, and finally plains which oscillated between poverty and wealth with bad management of the water supplies bringing malaria and excellent management the irrigation necessary for abundant harvests. Such striking contrasts within small areas naturally led to the movement of herds, people and property. The overpopulated hilly areas were willing to exchange wine and oil for the wheat they lacked and the higher areas were used as summer grazing for the herds that spent the winter on the plain. The mountain-dwellers followed a similar migration pattern either because they accompanied their herds or because they went to seek work in the towns and villages that were short of labour at the times when there was nothing to do at altitude. Distances were sufficiently small for thriving specializations to arise, even if they were never total, since it was always necessary to grow the greatest number of products destined for local consumption.

Landscape studies and in particular those devoted to rural areas highlight the many types of differentiation to be found there, but also demonstrate that these arose from different sets of circumstances and were not the main reason behind the formation of medium-sized territorial units or regions.

10.2 An Already Thriving Structural Hierarchy

The space occupied by traditional civilizations differed fundamentally from that of archaic societies by the nature of the networks which defined the former's social relations. A regular system of centres around which everything was organized replaced the expanding but sometimes quite distant links which remained unfocused, without a fixed hierarchy of central points.

Social relations had new institutional bases. The movement of goods was initially linked to the taxes raised by the civil authorities, the church and landowners. The taxes, tithes and rents collected in the farms or villages were rarely spent locally. Without military might the political powers would not command universal respect and they maintained this by directing the foodstuffs or monies levied from their subjects towards the camps, castles or fortified towns where their troops were stationed ready to intervene at any moment. Because money was easier to transport than foodstuffs it was conducive to the concentration and redistribution of means. It was in a prince's interest to check that the correct contributions were paid and not wasted in transit. He therefore ensured that everything that was owed to him passed through his coffers and then used it to pay for expenditure both in the capital and in the provinces. In order to achieve this he had to have at his disposal a hierarchized administrative system which covered the whole of the area over which he had control.

The church tithe followed more or less the same route, but was less systematically centralized, with Rome often leaving a considerable proportion to the diocese. Revenue from convents often took the form of income from land and, depending on whether it was feudal or market-based, resembled tax or income respectively. Where the tax was raised in kind the produce was stored in tithe barns, many of which still exist in the Paris Basin.

A portion of income from land ownership was spent locally by landowners; some lived on their estates all the year round while others lived there only during the harvest. A significant portion benefited the towns which they visited periodically to make purchases or in which they frequently lived. In the countryside in the West the proportion of income from land ownership spent in nearby towns and villages remained very high. It was a feudal tradition to live close to those under one's protection and this habit was perpetuated in the form at least of summer residence and, since land ownership was a sign of success, it was necessary to own a castle and to entertain there in order to become a part of the Establishment.

The income obtained was often supplemented by speculation. It was tempting to store the bulk of what was harvested in one's barns when a shortage was predicted and only to put it into circulation when a food shortage had set in. Whatever their origins, the sums accumulated promoted new trade routes which conferred an important role on markets in the new structures. Research carried out in Szechuan in China by Skinner revealed the role of periodic markets in the life of the countryside in the Far East where they set themselves

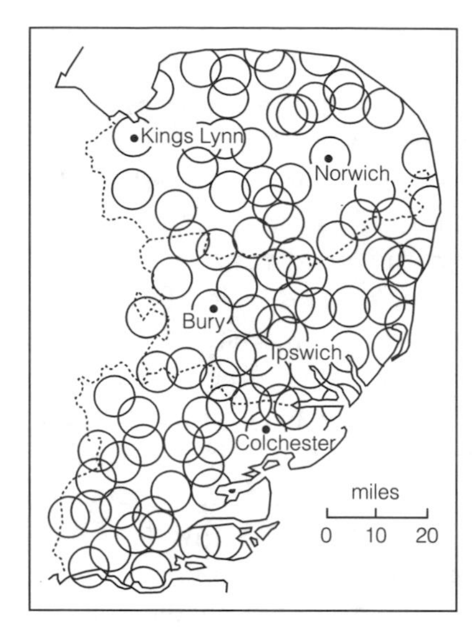

Figure 10.1 Rural markets in East Anglia in the sixteenth century

Established from the eleventh century onwards, the distribution of rural markets is regular except in the proximity of King's Lynn, Norwich, Bury and to a lesser extent Colchester and Ipswich. This gives an indication of the area covered by the 'oblique' ecological support of these centres. Movements were mostly contained within a four-mile radius.

Source: R. E. Dickinson, *City, Region and Regionalism*, London: Routledge and Kegan Paul, 1947, p. 81.

up in small villages and towns. In North Africa the souks were held for a long time in the middle of the countryside and their transformation into permanent central locations had not yet been fully achieved. By following the movement towards the creation of markets in Western Europe from the eleventh century onwards, one discovers the increasing role of commercial exchange in the basic levels of spatial organization (figure 10.1).

Hierarchical structuring played an essential role in the formation of medium-sized territorial units, but the forms it adopted varied greatly. In the West the grouping of areas of land was the work of either the feudal aristocracy, the church or the urban bourgeoisie.

The church in France had retained, since antiquity, the diocesan organization inherited from Diocletian's administrative reforms, thus providing a territorial model which affected the actions of feudal lords or of the sovereign. In Italy, according to the region, feudal hierarchies or constructs organized by the towns prevailed (figure 10.2). In the peripheral areas which benefited little from the rise in trade or craftsmanship such as Frioul or Piedmont the feudal world reigned supreme. Elsewhere towns quickly took over political control of the countryside surrounding them, forcing it to submit to their seigniory. Since many towns held the rank of earldom in the feudal hierarchy, the term *contado* was eventually used to designate rural zones.

In Islamic countries conditions differed from those prevailing in the Christian Mediterranean. Muslim land law contrasted with Roman or feudal law in that private land had no fixed status. Large farms were distributed by the sultan to his faithful, but these acts of largesse could be revoked at any time. Under these circumstances it was essential

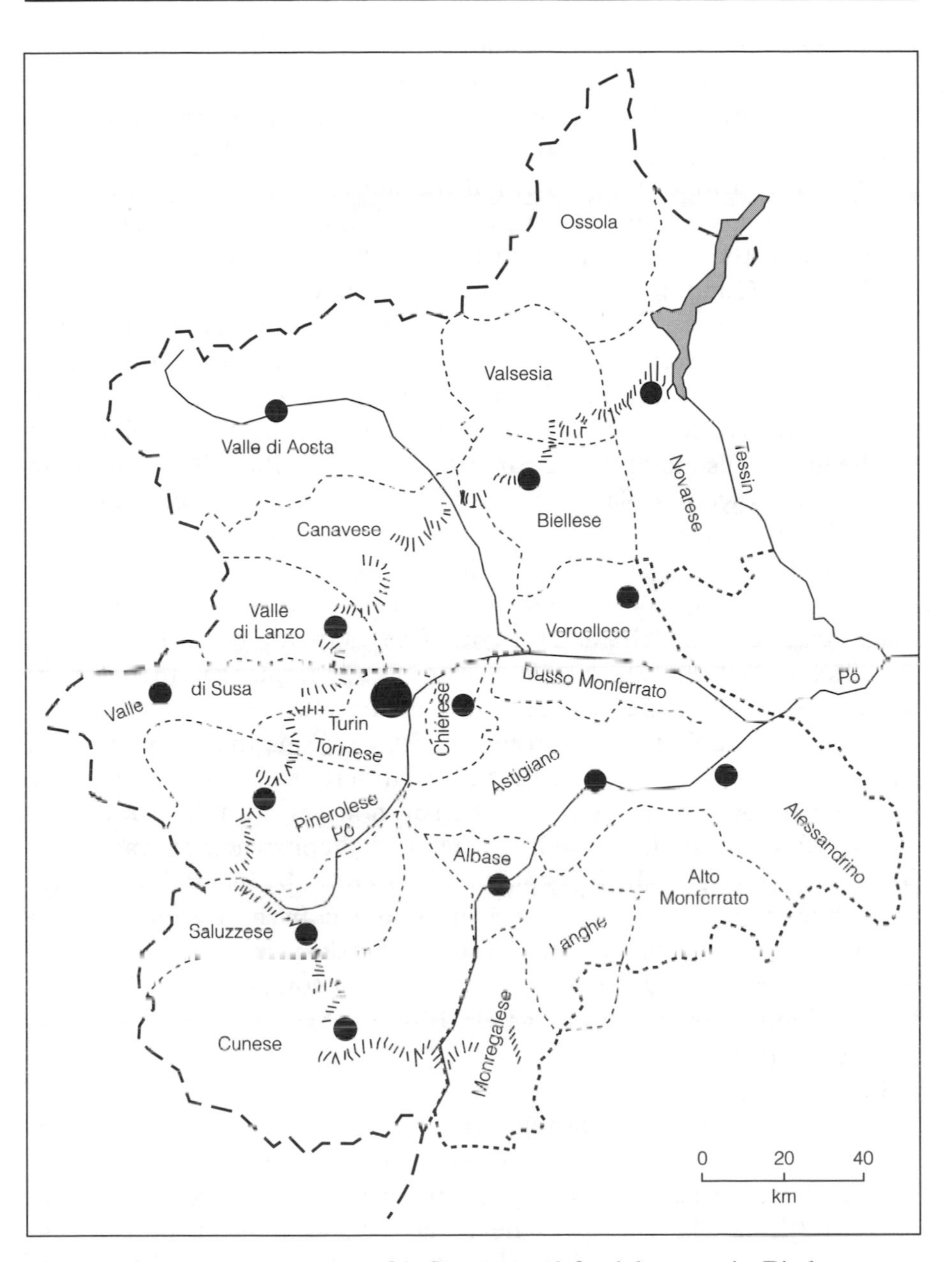

Figure 10.2 Urban sphere of influence and feudal power in Piedmont
The *pays* names portray the sphere of influence of towns in the lowland regions and the southern Alpine valleys, too restricted to form enclosed units. In the northern part of the Alps and in the hills throughout Asti, the place names reflect local autonomy and the weight of feudalism. The val d'Aoste, in spite of its name, belongs to this latter type.

not to be overlooked by the ruling power, and landowners only spent the amount of time on their estates essential for management and for the collection of taxes. Living in a nearby town held no advantages – it was necessary to live in the capital. These conditions were not conducive to the formation of well-developed hierarchical structures in the lower levels. Opportunities arose for a considerable number of towns when the leadership often collapsed, but instability tended to be the rule. Conditions changed from the sixteenth century onwards with the rise of the Ottoman administration in the West and of the Shi'ite religious clergy in the East. Administrative taxation in the former case and religious tax in the latter brought stability to the urban network and consolidated the regional level. In Iran, the regular visits by the great masters of the Koranic schools, Mechhed, Yazd, Ispahan and Qom, as well as Nadjaf and Karbala in Iraq, attracted wealth to the political centre, Tehran.

In the Chinese system the administrative principle had held sway since the origins of society and the towns of northern China still owe their structure and character to it. The network of administrative centres for each prefecture or *hsien* was already in place in the north of the country and in the central basins in the Han era. Actual centres appeared later only in the northern coastal regions the topography of which was seriously affected by the meandering of rivers and in the hilly and mountainous regions of the south as yet not fully populated.

The often remarkable stability of territorial constructs in traditional societies could generally be explained, therefore, by hierarchical structures linked to the ruling power or to the church. Certain French provinces have the same boundaries as the diocese from which they emerged, which in turn correspond to Gallo-Roman cities, that is to say, to Gaulish tribal groupings established for the most part in the third century BC.

In areas in which the churches had difficulty in establishing their authority the classic type of regional organization lacked stability, remained fluid and did not cover the whole area. Gilles Sautter[2] observed that in Black Africa or even in the Sudan or on the high plateaux of Eastern Africa, the state was never able to establish a power base which guaranteed it success and permanence. As a result the regions of traditional Africa have three specific features:

> Firstly, their most general feature is *isolation*. They appear in every respect to be like islands: islands of population and organization in the middle of areas without any structure. . . . The second feature of traditional regions in Africa which is a corollary of the first is the *absence or weakness of organic links on the level of inter-regional structures*. . . . The third and final characteristic common to all traditional African regions

is consequently *their extremely uneven geographical size.* (Sautter, 1968, pp. 96–7)

In societies the structure of which remained incomplete even when institutionalized links were already beginning to diversify, regional construction beyond the type of mutual awareness and familiarity described by Roland Pourtier remained irregular and often original in its configuration and could not cover the whole area. The specific types of territorial organization found in intermediate societies never gained control in Black Africa.

10.3 A Strong Cultural Component

The regional differentiation which was very apparent in traditional civilizations was often expressed as much through cultural characteristics as through the landscape and in the former case through details which had no functional significance. The geography of dialects in France is founded on that of the diocese, that is, on the oldest territorial constructs. In eras where withdrawal into the local area was the most marked, the priest was part of a network in which there was constant movement. The priest went on retreat or to the bishopric; the bishop toured his diocese and monks belonging to certain orders were highly mobile. These interactions helped to maintain a certain linguistic coherence.

Cultural diversity was very evident in costume, architecture and tools. Today the Quercy and the Périgord conjure up visions of the pigeon lofts to which the façades of many houses owe their noble appearance and the stereotypical image of Alsace is of half-timbered houses with large steeply sloping roofs. In the Touraine the proportions and quality of architectural detail in the large houses made of calcareous tufa are striking and reveal the influence of an ancient urban civilization. Differences of this type appear to be so natural that they are considered to be an integral part of the traditional world. For many French people today a region only exists if it has its own original cuisine and typical style of architecture. In the last century, traditional costume, which has since disappeared, could also have been included in this list.

Studies by folklore specialists both in Great Britain and in France have demonstrated that differentiation certainly has ancient roots, but that it was greatly accentuated between the end of the eighteenth century and 1880. At a time when increased mobility was breaking down the old territorial identities, it demonstrated the powerful reactions of well-established societies. As travel became easier so people discovered the extent to which the signs which distinguished people and places helped them to retain a strong sense of local or regional uniqueness. The notion that a Breton woman wore a white starched headdress became a stereotype throughout France; however, each canton within Brittany affirmed its own identity by creating its own unique design.

Local and regional differentiation in cultural traditions was very real in traditional societies, but should not be overstated. Evidence provided by ancient habitat or by what folklore museums know about the past is often only a century and a half old. Landscapes, like the habits and the pattern of life, were doubtless more uniform in Europe at the end of the Middle Ages or at the beginning of the modern era than is generally thought.

The fundamental cultural dimension of territorial constructs of the traditional world had other roots. It stemmed firstly from the architecture of the institutions proper to each area of civilization. The apparent absence in traditional India of the type of regional structure generally found at this level of development seems surprising, but the training and interests of British and Indian geographers to whom we owe our conception of the country certainly bear much of the responsibility. The former set greater store by close analysis of the statistics supplied by the wonderful *Census of India* than by contact with the population. The latter then also adopted this type of approach, but the cultural atmosphere in which they were immersed, moreover, led them to attach greater importance to castes than to regional units as a framework for the development of identities. They were not entirely mistaken: the sense of belonging that developed in the Indian world differed radically from that of the Melanesian archipelagos, previously discussed, in that we are no longer dealing with a 'geographical' society. However, careful observation does yield evidence of medium-sized territorial structures – *pays* and region – and this has always been the view of French geographers. David Sopher[3] has tried to specify the origins of these differences and the points at which resemblances with other cultural areas appear.

The division into castes and its concrete translation into *jati*, together with the exogamic structure of the sub-groups of which they were composed, formed more widespread links of kinship right across India. Among the members of one's *jati*, one would find in one's own village, and in those within a radius of 20–30 kilometres or more, only girls whom one could not marry because they belonged to the same clan. It was frequently necessary to go further afield, often 50 or 60 kilometres, to find a wife. This implied links over a medium distance and created a cultural permeability within each sub-caste which was not found anywhere else. Political structures were not well developed and lacked stability. The diversity of the groups meant that the administrative districts were less conducive to fostering a sense of identity than elsewhere. However, in this vertically divided society religion provided a horizontal structure and it was often by visiting the same holy places and by going on pilgrimages that people from an area discovered what they had in common.

The variety of ontological status conferred on places was decisive in the structuring of areas. This phenomenon obviously existed at the level of elementary units and often had its roots in very ancient traditions. The network of sacred sites of Catholic parishes generally merely adopted that of the pagan cults it replaced. It was in order to obliterate these that the cross was imposed on everything deemed to have a numinous power.

Most rural civilizations had a very clearly defined religious topology. Philippe Pelletier[4] points out that in Japan it is primarily the mountain 'which symbolizes the expression of the creative force; it personifies a *kami*, a divinity, or spirit. It is a first principle, a point of departure.' He goes on to state that 'therefore, there is a religious axis pointing downwards towards human settlements, that is towards the base which represents the voyage from spring to autumn of a *kami* who personified the mountain for the people' (Pelletier, 1987, p. 84). Shinto shrines lie along this axis from the mountains or valley floor, from the village in the hills at the base of the mountains through which the plain was populated and from the plain itself.

Cultural facts are equally necessary to explain the rise of larger-scale territorial structures. François de Polignac[5] recently reinterpreted the formation of the Greek city in this light:

> The State in the 'dark ages' seemed to have been a horizontal juxtaposition, where in the case of low population density, ethnic, tribal and geographical divides kept the various social entities (market towns and villages built around several noble houses) in a state of relative autonomy within a limited sphere of activity, trade and influence. (Polignac, 1984, p. 153)

There was however pressure to regroup and to create new structures and it was through religion that this pressure found its expression and that the landscape was restructured:

> The flowering of religion which occurred at the end of the Geometric period demonstrated . . . that the cultural factor was the only one affecting society as a whole. This evolution was in fact a sign of the formation of a society which seemed to be becoming self-aware by reappropriating the past and consecrating it, by gradually redirecting the offerings made

> to burial sites towards the shrines to the gods in which a ritual transformed a collection of individuals or groups into a community belonging to the same sect. Thus it was through religion that a social body was gradually born founded on a 'religious territory', i.e. not a symbolic form of the real landscape, but rather a way of viewing cohesion in a given area. (Polignac, 1984, pp. 154–5)

After the Geometric period the city was born from a double spatial polarization:

> The fact that this area . . . was frequently bipolar in structure with a society defining and organizing itself simultaneously about its centre and on its geographical periphery was neither unusual nor arbitrary. It was because exclusive centralization required a submissive attitude on the part of the peripheral areas that the cults practised in the centre proved to be inadequate to the task of forming and creating the consensus and rejections required to build a unified and viable society over a whole area. The major cult of the 'territory' was the pole for the social constitution of the city as is demonstrated by the centrifugal wave which drew the whole of the population to the fringe of this area which was the median point of society. The first movement, however, was accompanied by a second which was the formation of a collective power structure in the centre, the concrete result of the unification of local aristocracies into a ruling body. . . . The central pole (which was to become the urban pole of the Greek world) by virtue of worship both of gods and heroes was the political foundation of the city formed by the centripetal movements of the members of its assemblies and councils. (Polignac, 1984, pp. 155–6)

François de Polignac's analysis clearly demonstrates how the birth of the 'region' as a reality, i.e., a medium-sized territorial organization and in this case the city, used the ontologization of the space to cross the threshold towards which social evolution was tending, but which was proving to be a stumbling block. It is, therefore, important not to separate the study of symbolic models from analysis of overall social and economic forces at work within an area.

At a higher level the role of the sacred is equally apparent. Political constructs all verged on the religious and this was particularly true when traditional societies were being formed.

In a country like China the Emperor was revered for the inestimable service he performed in periodically reconciling society with the fertility gods and banishing evil forces. To perform these rites he had to be at the centre of the world, at the point at which the cosmic axis connects the earth to the powers in the heavens or the underworld. All the symbolism of the 'pivot of the four corners'[6] on which the imperial city was designed lies here. It was a tangible manifestation of the geographical privilege which conferred meaning on the Emperor's religious acts. The plan of the capital towns of the prefecture was based on those of the capital city and served throughout China as a reminder of the basis of the justification for imperial power. Research carried out by Joseph Rykwert[7] into Greek and Roman geometric plans and into the role of the axis of the world and the *mundus* demonstrates that the western tradition differed less than might have been thought from eastern rituals.

10.4 The Regional Organization of Nations and Major Areas

In historical civilizations the nature of economic life and trade meant that regions were often an ideal framework for the type of specialization that was possible at that time. The complementary nature of these specializations was sometimes conducive to the transformation of regions into small states. Lucien Febvre wrote of sixteenth century Franche-Comté:

> This extreme variety within Franche-Comté helps us to understand to a large extent the enduring autonomy and lasting vitality of the state. Nowhere else says Gollut were there so many commodities to be found in such a small area. . . . Was Franche-Comté not rich in cereals – wheat on the plain, rye in the mountains, barley and millet? Were the white wines not served at the tables of Erasmus, François I or Emperor Charles a source of fame and pride throughout the world? . . .
>
> The communal enjoyment of such a rich heritage was an ideal way to forge new and powerful bonds between the people of Franche-Comté in an era when travel was difficult, land was being parcelled and jealous protectionism was the order of the day. Natural and human forces could be seen to be working together. However, contrary to appearances, man's contribution was the larger of the two and it was he who definitively formed a political unit, a state from the disparate elements. (Febvre, 1970, pp. 29–31)

This interplay of complementary economic factors explains to a large extent the viability of the regions in the organization of land before the Industrial Revolution, for it was in the heart of the regions that trade was organized. The Mediterranean world offers many illustrations. It was to the advantage of all the environments that made up this world to participate in the construction that facilitated the safe exchange of goods and travel – mountains, hills, plateaux, piedmonts, fertile basins, irrigated plains, disease-ridden marshes and shorelines stimulated by trade and fishing, but suffering on account of their small area. Greek cities were already reaping the benefits of the complementary relationship between the forest and maquis of the high ground which supplied wood, coal, honey and livestock products, the hills given over to producing the three famous Mediterranean products wheat, wine and olive oil, and the shoreline that gave access to foreign trade. Closer to home regions such as Tuscany or Provence displayed similar complementarities over larger areas.

Conditions in the Mediterranean world were particularly conducive to trade on account of the sea and the large, dry plateaux over which mule trains in the north and camel trains in the south and east could travel easily. These complementarities were often exploited within the framework of larger political constructs or even on the scale of true international specialization. Catalonia was a good example of the former type. It was already more than a region by virtue of its size, but the solidarity, which existed between the mountains with their herds, wood and iron, and the dry interior, which supplied wool from its sheep, oil from the Borjas blancas and salt from Cardona and the network of massifs and coastal plains, the climate and soils of which favoured the cultivation of either wheat, maize or vines, created a highly integrated economic unit. It was not a closed unit, however, and Barcelona received a proportion of its foodstuffs from the Balearic islands, Provence, Sardinia and Sicily.

These conditions led to the creation of political units that spanned seas. Greek and Phoenician colonization in the period of the founding of cities was a precursor of this evolution. During the Middle Ages experiments of this type proliferated of which the Anglo-French state spanning the English Channel at one time is an example. Conditions in the Mediterranean were even more conducive to constructs of this sort. At one stage the Catalan state encompassed a considerable part of the western Mediterranean coastline. Genoa and Venice were good examples of maritime city states the trading posts and colonies of which consolidated their key trade links from the Crimean coast or the Levant to the seas surrounding Italy.

National states were generally too huge and the conditions for communication too difficult for economic specialization to give rise to a contrasting regional organization. Most exchanges remained on a local or medium scale even if they were often outside the administrative framework. It was only along the navigable rivers and sea coasts (figure 10.3) that the situation changed. Transport by waterways was cheaper than transport overland and was conducive to greater specialization than in areas where the continental mass widened, but this specialization was never total. From a contemporary perspective, these states appear to be somewhat strange, since peripheral provinces traded with distant countries via rivers and seas rather than within their national

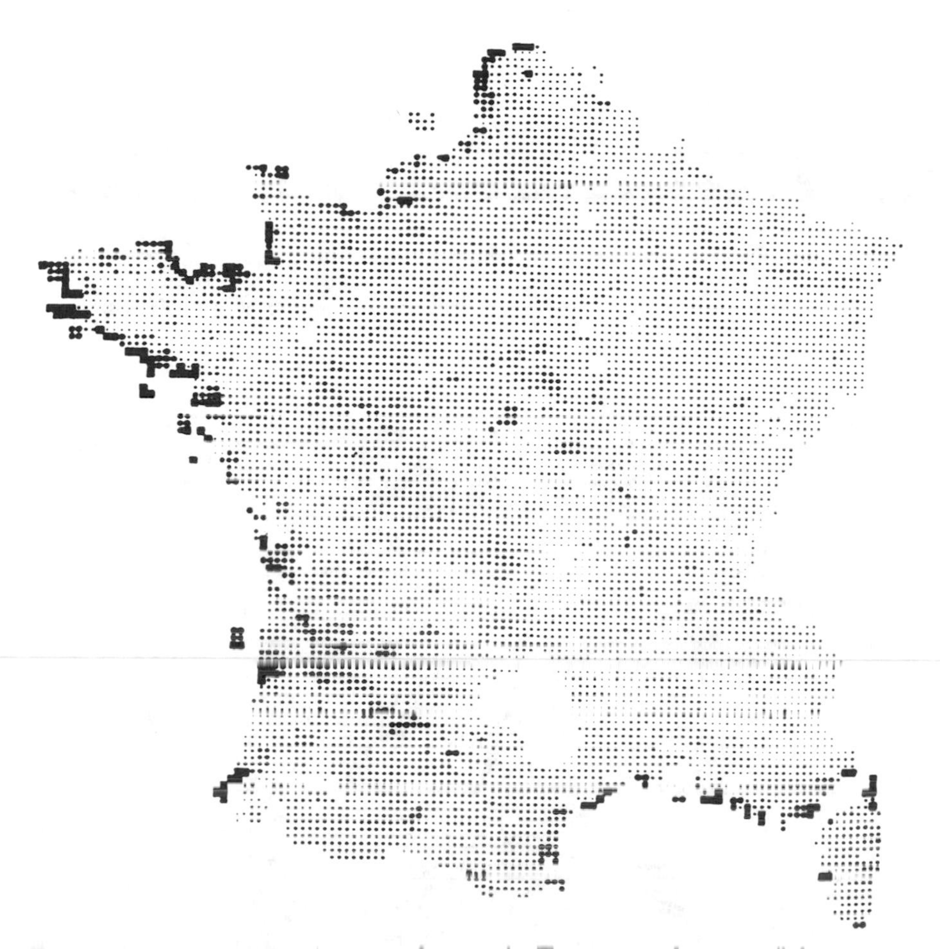

Figure 10.3 The opening up of space in France on the eve of the industrial revolution: the importance of maritime factors and, in a secondary context, fluvial factors

The analysis of the occupations of French military conscripts for the period 1819–30, underlines, as a secondary factor, the contrast between the interior space of France and that turned towards the exterior.

Source: M. Demonet, P. Dumont, E. Le Roy Ladurie, 'Anthropologie de la Jeunesse masculine en France (1819–1830)', *Annales, Economies, Sociétés, Civilisations*, 31, 4, 1976, p. 757.

borders. In France, Provence and the Languedoc looked towards the eastern Mediterranean for outlets for their goods.

Lyon had a similar orientation on account of the Rhône valley and traded with northern Italy via the Alpine passes. Bordeaux sold its wine throughout northern Europe and grew wealthy on the trading links forged with Africa and the West Indies from the seventeenth century onwards. The valleys of the Garonne and the Dordogne provided it with an export base in the form of grain, flours (hard wheat),

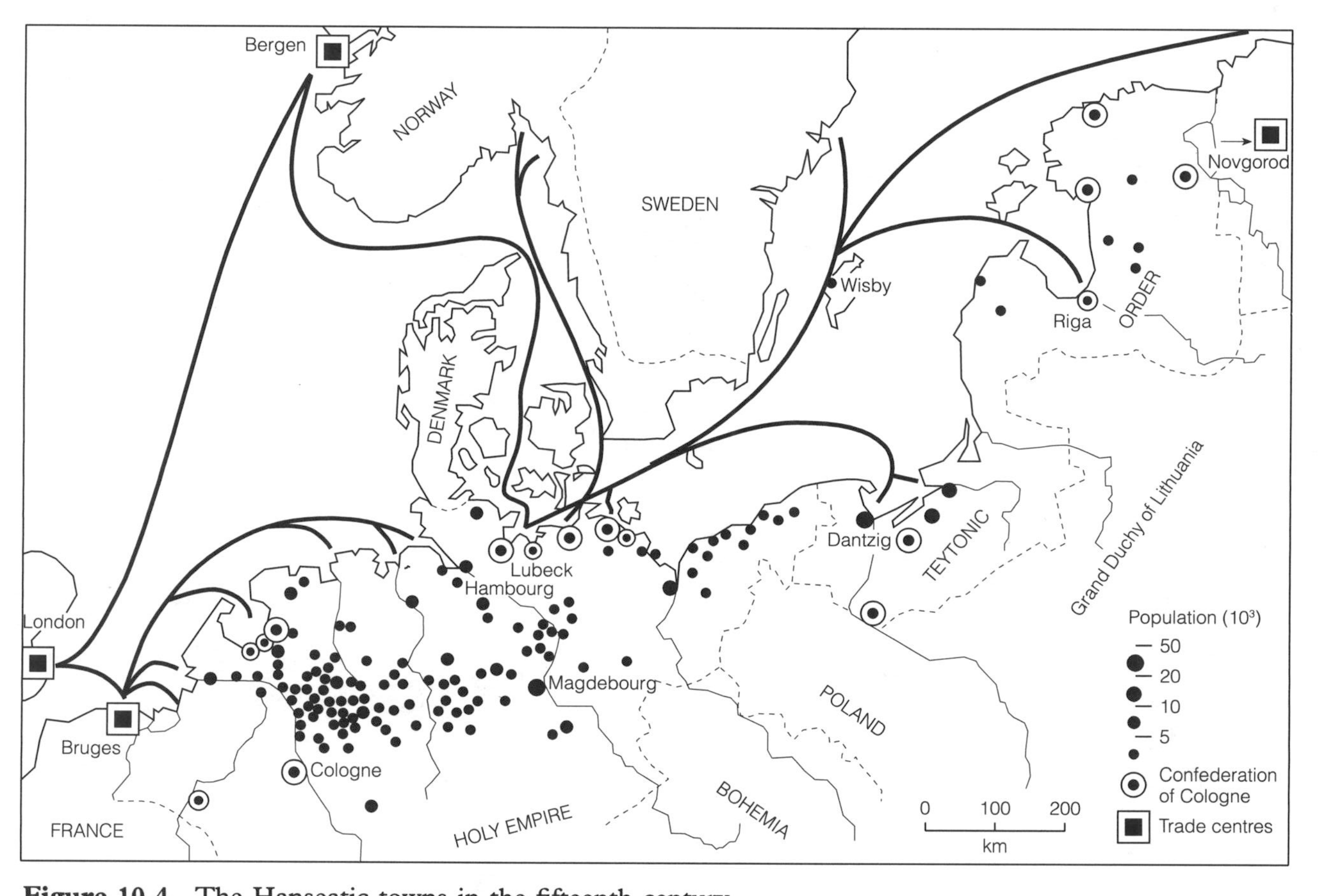

Figure 10.4 The Hanseatic towns in the fifteenth century

Source: Grosse historischer Weltatlas, Bayerischer Schulbuch-Verlag, 1970, t. II, p. 123.

wines and dried fruit from a large proportion of the Aquitaine and the fringes of the Massif Central. Nantes, which had the same orientation, benefited from having as its hinterland the Loire basin, the best river for navigation in the prevailing conditions.

It would be incorrect to assume that waterways stimulated the whole of the French interior. According to Vauban, who was very familiar with the country, they had a real impact along corridors two leagues wide. Regions integrated into long-distance trade covered a tiny proportion of France's national territory under the *ancien régime*.

Long-distance trade was organized and controlled by the ports through which bankers, merchants and sailors had forged distant ties, had found reliable sources of luxury goods and had opened up markets. These cities exerted an influence on the life of the period that was disproportionate to the size of their populations. They did occasionally guarantee the prosperity of the political structures which encompassed them as was the case in Flanders or a little later in Holland. However political forces in large states, as in the case of France, were too indifferent to their needs for large trading posts truly to flourish. Often, as in Italy, centres of foreign trade were sufficiently large to organize themselves into city states. In northern Europe the Hanseatic towns formed a league of city states which therefore thwarted the ambitions of the land-based powers surrounding them (figure 10.4).

The organization of space into large units was, however, linked much less to the economic forces than to the interplay of cultural factors and to the transformation of the military techniques or ideologies on which powers rested. Empires were successful when one group gained military superiority and backed it up with a religion or philosophies that favoured acceptance of a universal sovereign. In Europe, the image of imperial Rome still fired the imagination but ethnic diversity led to the formation of other political structures. In the West and North at least, states relied very early on the emergence of a national consciousness. However, other forms of organization were possible as was demonstrated by Central Europe.

The founding of modern states was accompanied by the concentration of power and its increasing geographical stability. It was not possible to control a huge area without a bureaucracy and the latter cannot function unless it is structured around a major pole. Among the often undifferentiated mass of small local and regional units the national space was therefore marked by the emergence of capitals, which gave rise to distinctive conditions in the surrounding areas. They collected revenue throughout the country so that their economies often became totally monetarized very early. The needs of an already large population that was richer than elsewhere stimulated production and specialization in all the regions which were linked to them. The Seine valley, and those of its tributaries, played a decisive role in the provision of food supplies to Paris, which explains the early appearance of speculative vineyards in lower Burgundy, the intensive exploitation of forests for firewood in the Morvan and the prosperity of cereal cultivation even at a great distance from the capital. Herds from Normandy or the eastern provinces converged on the markets at Passy and Sceaux. A whole geography of cheese grew up based on the Parisian market – soft cheese that did not keep well was made near the capital while cheese that kept well was made 20–25 kilometres away.

In Japan under the Tokugawas between the eighteenth and nineteenth centuries the rise of Edo (modern-day Tokyo) – the shogun's place of residence – was such a drain on revenue that the whole nation was affected. Coastal shipping trade brought into the capital the rice necessary to feed the capital and the cotton to clothe it from all over the country.

The adminstrative authorities established in the towns chosen by the central power often had a greater jurisdiction than those in the provinces. Urban development was therefore halted within a radius of 100–200 kilometres or more. The social pyramid in small and medium-sized towns was less varied. Thus as early as the sixteenth century the preconditions existed for what would later be referred to as the concept of 'Paris versus the French desert'.*

In traditional civilizations, forms of spatial organization were very different from those which gave rise to the industrial and transport revolutions and as a consequence are often misunderstood. The study of these forms is, however, fascinating in itself, but it is essential to an understanding of how regions evolved in the last two centuries, since territorial structures remained static even when conditions changed.

Further Reading

Mair, L., *African Kingdoms*, London: Oxford University Press, 1971, 151 pp.

Skinner, G. W., Marketing and social structures in rural China, *Journal of Asian Studies*, vol. 24, 1964, pp. 3–35, and vol. 25, pp. 195–228.

Tuan, Y.-F., *China*, London: Longman, 1970, 225 pp.

Wallerstein, I., *The Modern World System*, New York: Academic Press, vol. 1, 1974, 410 pp., and vol. 2, 1980, 370 pp.

Wolf, E.R., *Peasants*, Englewood Cliffs, NJ: Prentice-Hall, 1966, 116 pp.

* A phrase coined by J.-F. Gravier in his book, *Paris et le désert français* (Paris: le Portulan, 1947), alluding to the over-concentration of activity and power in Paris and the relative economic, political and demographic weakness of provincial France.

11

Forms of Regional Organization Stemming From the Industrial Revolution

The industrial revolution and the transformations linked to it in the domain of agriculture and transport disrupted the relationships between human groups and space. The innovations which ceaselessly succeeded each other over the space of two centuries, involving all aspects of production, modifying the level and type of consumption, changed the conditions of transmission of knowledge and styles of culture and were set in a general movement of symbolic thought. When one reflects on these changes, there is a danger of getting lost in the detail: to avoid this it is as well to step back to define long-term trends.

The changes which so profoundly modified the relationships between space and human societies were set in a fourfold progression:

1. the recourse to concentrated forms of energy inflated the link between resources, work and population;
2. the mobility of goods, people and ideas heightened competition, reinforced the role of routeways, nodes and networks in spatial organization and enlarged the frameworks of life;
3. environmental constraints did not disappear; they were moved since the impacts springing from the concentration of population and activities disturbed the ecological pyramid over ever wider areas;
4. the birth of mass cultures and the explosion of educated cultures were the first steps in an evolution marked by the generalization of public education, the progressive erosion of traditional cultures, the secularization of space and the growth of new forms of what is held to be sacred.

The evolution, which began in the last decades of the eighteenth century never stopped, but the forms of territorial organization successively assumed configurations that remained stable over relatively long periods. We may distinguish two major stages, the industrial phase and the post-industrial phase. Two models of spatial structuring characterize the first: the British model, and the centre–periphery model, which finds its clearest expression in the United States.

11.1 Territorial Organization in the Age of the Industrial Revolution: the British Model

The impact of the industrial revolution

It is in England, where the industrial revolution began, that it is easiest to grasp its initial features. International trade prospered in proportion to imperial expansion and the capture of new markets. An unprecedented demographic expansion began: the cause (it supplied labour) and consequence of imperial expansion (it was easy to emigrate), a more secure food supply (famines and shortages disappeared) and the opening of manufacturing which offered employment close by. Mechanization commenced in textiles thanks to new looms and spinning machines, taking advantage of improvements in water power and steam engines, but other sectors of the economy were not affected initially. Agriculture improved due to more rational rotations and a closer integration with livestock, but remained rather unproductive and employed many hands; transport remained relatively slow and onerous and traditional forms of the transmission of culture persisted.

The new features were obviously related to the birth of industrial regions. This was a consequence of the impossibility to transport the new forms of energy being exploited: manufacturing formed ribbons along swift flowing rivers with a regular discharge, or agglomerated at pitheads (the railways only appeared forty years after the beginning of the changes). Thus, 'black countries' were formed, quickly affected by industrial pollution, aggravated in some cases by relief favouring temperature inversions. In the winter, fogs were impregnated with poisonous gases while dirt particles encouraged their occurrence. Smog replaced fog, at times provoking a dramatic increase in pulmonary diseases.

Manufactured goods replaced those previously supplied by artisans: they were produced for current consumption. The production chains remained short, including the case of textiles where spinning, weaving, finishing and dyeing continued to function as distinct stages. Even where they were carried out in separate factories or firms, it was imperative to be located close to each other to facilitate exchanges of information, still very important at these stages. Thus, over small areas accumulation of complementary activities took shape. At the end of the nineteenth century, Alfred Marshall coined the term 'industrial district' for this type of manufacturing zone. In Lancashire, each locality specialized in spinning (in the south, around Bolton and Oldham, with medium grades in the latter town, at Ashton and Middleton, and for fine grades at Bolton, Chorley and Preston) and in bleaching with Bolton as the main centre. The chemical products required for these operations came from Widnes and St Helens; textile machinery came from workshops in Oldham, Bolton, Blackburn and Burnley. All these activities were due to the proximity to Liverpool, the market for imported fibres, to acquire cotton cheaply, to access the largest selection of quality and to export the final product efficiently. Manchester served intermediate transfers: its stock exchange fixed the price of yarn and cloth. This is where dealers operated, mainly exporters, who undertook responsibility for manufacturers' contracts with customers. Birmingham and its surrounding towns offer another example of an industrial district: here we find the entire cycle of ironworking. It was from here that almost all English production of bars, nails and rings came, as well as arms, knives, padlocks and locks. Engineering expanded early – it was here that Watt manufactured the first steam engines. Manufacturing production expanded but the construction goods sector still had only a limited market (except for transport, where the needs related to the construction of railway lines after 1840, and shipbuilding after 1870 provided a sustained demand).

Increased production resulting from the birth of manufacturing is only possible by enlarging markets: the internal market can be supplied but exporting is necessary to absorb production, which was rapidly increasing without really diversifying. The nation did not as yet constitute the fundamental framework for growth. Central locations in relation to the national space did not bring advantages to the industrialists. To be successful it was better to locate in ports; in the absence of ports, the most suitable sites were distributed along waterways permitting cheap transport of production. The English model at this stage did not show the opposition between the centre and periphery of the national territory. The new industrial centres were determined by the coal deposits in the hercynian massifs and their concealed extensions at shallow depth. The Midlands coalfield, where the industrial revolution was born and which is the best situated to serve the national market due to its central position, did not experience such spectacular growth as the coalfields of Lancashire, Yorkshire, Northumberland, South Wales or the lowlands of Scotland: located in coastal zones, exportation was easier there (figure 11.1).

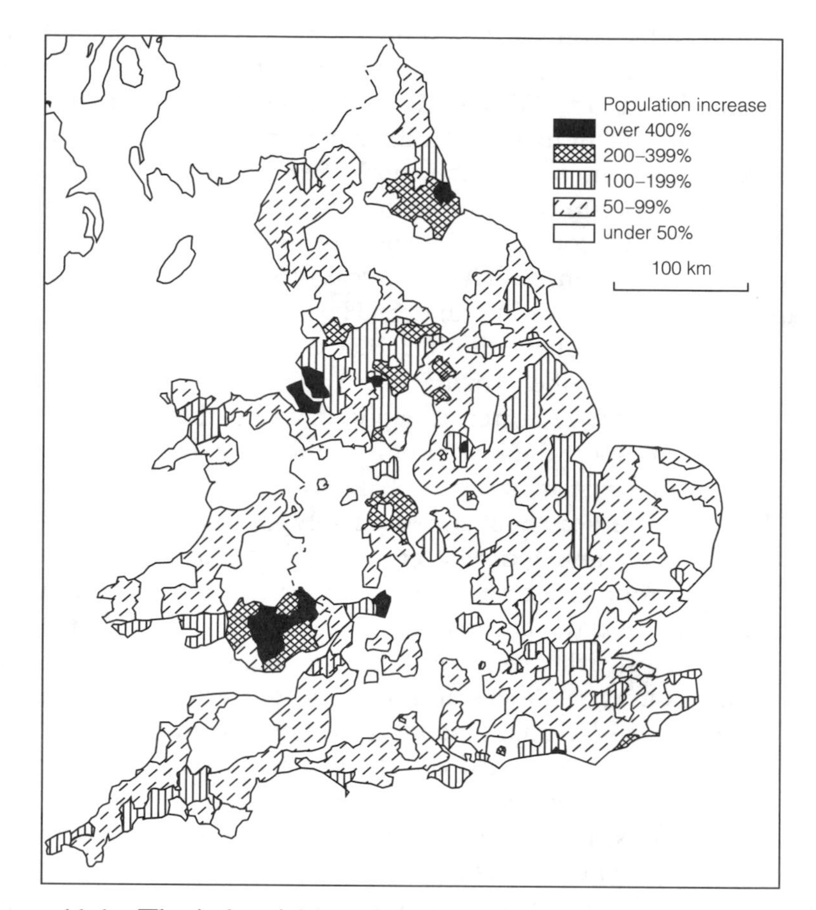

Figure 11.1 The industrial heartlands in England and Wales in the mid-nineteenth century as revealed by demographic change between 1801 and 1851

The England in rapid expansion was that of the littoral coalfields, or those close to the coast, from Northumberland to South Wales, passing through Yorkshire and Lancashire. The Midlands and London were growing less rapidly.

Source: J. W. Watson and J. B. Sissons (eds), *The British Isles: A Systematic Geography*, London, 1964, p. 232.

The new forms of territorial organization

The nation could only exist as an economic unit by virtue of financial markets, and of networks to collect savings, which were established from London, and through the application of customs protection for agriculture. Rural densities remained high since the labour required stayed constant per unit of surface area, or increased due to intensification, but commercialization made rapid strides as it was stimulated by new markets created by industrial concentrations. Specialization of production asserted itself, even though it was still handicapped by transport costs. The old centres in the urban network benefited from the evolution since they saw their commercial functions and their role in the new capital and money markets increase. Regional sentiment, born of a long history was still universally felt, but zones of speculative farming and monoproduction extended without reference to traditional divisions. Regional specialization, which had already become strong in the case of industry, became established in rural areas. The urban hierarchy became more emphasized: the towns had to meet a greater and more diverse demand for services because production became more varied and consumption levels began to increase. More or less everywhere, the hinterlands of important towns strengthened: polarized regions began to play an important role. Regional capitals were born. In predominantly rural regions they were regularly distributed and evenly spaced, since to prosper they needed to serve a quite large population of average density.

In the South-West (figure 11.2), which escaped industrialization (and even saw disappear, in the Cotswold villages, artisan workshops that worked wool until the invention of new looms), rural intensification enriched the fertile countryside of Devon, Somerset and, further north, the counties of Avon, Hereford, Gloucester and Worcester (using the current nomenclature of the counties). The towns, historically prosperous, remained animated, but only developed modestly. The important markets, bishoprics generally, drew part of their diversification from their commercial functions, as evidenced by Exeter, Worcester, Gloucester or Hereford – but not all achieved a functional conversion, as Wells, which stagnated, shows. The growth of these centres is at any rate limited. They were challenged by a new generation of middle-sized towns. Some owed their prosperity to the railway, as in the case of Swindon. The majority had a tourist function, which prospered both in spas (Bath and Cheltenham, very fashionable in the eighteenth and at the beginning of the nineteenth centuries respectively, benefited from outstanding urban design) and at the coast (Bournemouth, Weymouth, Torquay or Barnstaple may be cited). A single centre dominated fully: Bristol. Its growth goes back in history, linked to the wool trade since the Middle Ages, then to links with the Antilles and America since the beginning of the seventeenth century. A city of ship owners, she secured industries related to imports of colonial products, sugar or tobacco. But its fortune was also based on regional functions: Bristol became the capital of the west and controlled the highly diversified countryside surrounding it. The proximity of London, and the pattern of railway lines focusing on the capital rather than Bristol, limited this role however.

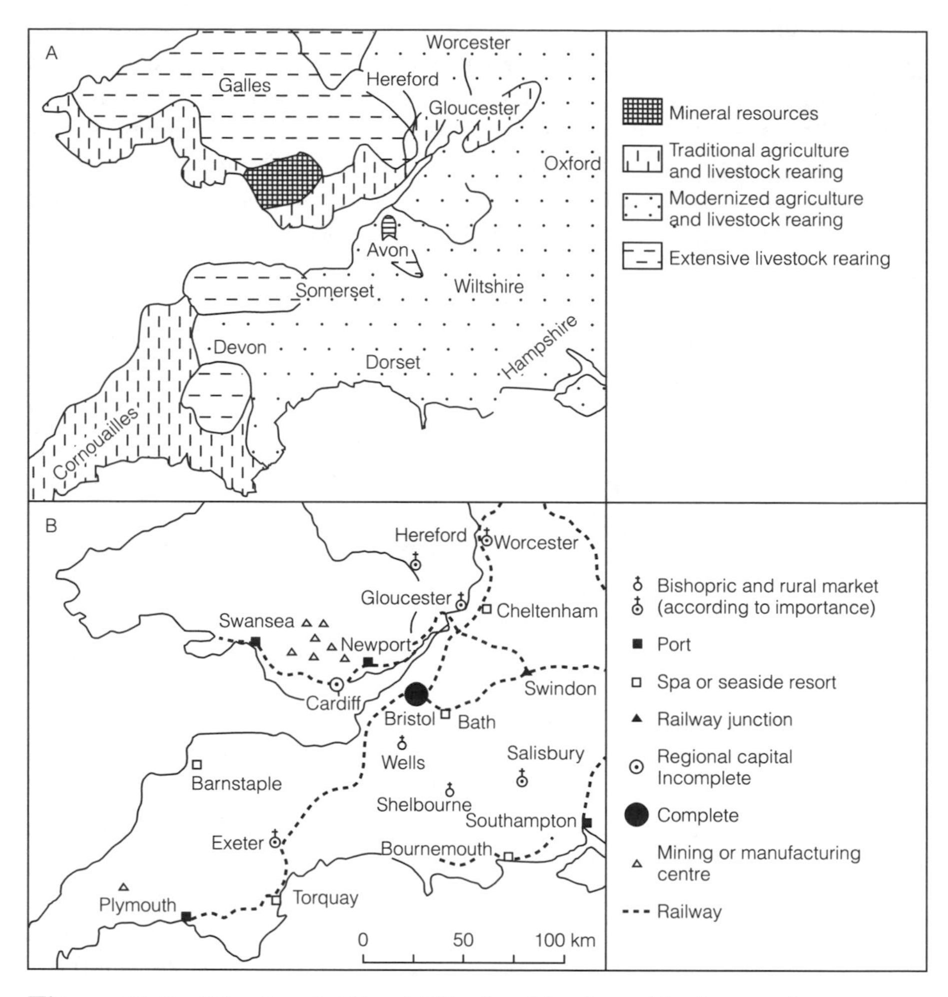

Figure 11.2 The South-West of England in the mid-nineteenth century

A In contrast to the agriculture and extensive animal rearing prevalent in Wales and Cornwall, the South-West enjoys a rich agriculture. Its only mineral resources are the small Bristol coal basin and Cornish tin, and the coal of South Wales, which, isolated from Bristol by the Severn Valley cannot compensate for the absence of local resources.

B In contrast to the South Wales coalfield, for which Cardiff plays the role of an incomplete regional capital, South-West England is structured by a network of towns (bishoprics and market towns) that developed slowly through a new layer of seaside resorts or rail towns, and organized by a capital, Bristol, which unfortunately did not manage to achieve a railway hub to counter the influence of Birmingham upon Worcester and of London upon Somerset and Devon.

In the new industrial concentrations, towns were much closer to each other. The mesh of towns, for a long time relatively regular in rural zones, became more complex in urban areas. Patrick Geddes coined the term conurbation for them: one no longer found the classic distinction between town and countryside in districts of massive industrial accumulation, in Lancashire for example. The urban built-up area was ubiquitous. The centres which played a metropolitan role were hard to distinguish on the map from those with purely manufacturing functions. However, they prevailed by virtue of their functions: Liverpool and Manchester scarcely had factories; these towns lived from the organization of production circuits and from commercial activities. They were the equivalent in the industrial world of the regional capitals in the predominantly rural areas.

The nation which experienced the first phases of the industrial revolution has thus retained many traditional features:

1 the absence of being inward-looking;
2 the importance for manufacturing of littoral or sub-littoral locations, the only ones capable of opening onto international markets;
3 the large significance of agricultural employment, which explains the persistence of predominantly rural polarized regions.

Spatial organization assumed new forms to the extent that it was marked by:

1 the birth of industrial regions which disturbed the regularity of traditional regional outlines;
2 the increased polarization of rural areas and by the differentiation of their productive base;
3 the multiplication of the economic functions of the capital, where stock exchanges and banks employed more and more people.

New forms of dedicated land use for purposes of recreation and consumption became defined for the benefit of the wealthy classes: tourism appeared in a modern form, swelled the clientele of the spas and gave birth to narrow ribbons of urban life along the coasts.

The uniformization of space, which began to be defined, and, in parallel, a secularization linked to the rationalization of thought and ideologies, did not provoke very brutal reactions. Rejections are evident, however: they are perceived in rural regions which resisted generalized commercialization and remained loyal to a self-sufficient polyculture that guaranteed a real independence to small-scale farmers. In England, where the social structure was uniform and where the land was in the hands of major landlords, such changes were scarcely possible. They were observable by contrast in Ireland or in the case of the Scottish crofters. The rejection of homogenization was doubtless stronger on the continent, but even in Great Britain, the paradoxical consequence of the construction of a unified market was for certain groups to assert their identity by insisting on what made their material culture distinctive: feelings of identity persisted for some time.

The English model is limited in nature. Since growth there was still based on a narrow range of industrial production and a national demand, which of necessity was inelastic, an incessant enlargement of export markets was implicit. Economic rivalries were affirmed in proportion to the adoption of English innovations by European and North American industrialists: competition hardened since the global volume of the market scarcely increased.

The contagion of progress and new countries

When the industrial revolution took off in England it was not surprising that others were unaware of the advance enjoyed by this country: it took time for innovations to diffuse. The advantage that Great Britain enjoyed was real, but it is in the nature of things to see it progressively eroded by the adoption elsewhere of the technology she had created. Efforts were made on the British side to retain the monopoly of new manufacturers, but this conflicted with the interests of machine manufacturers tempted by exports and with the mobility of engineers, foremen and skilled workers attracted by offers of high wages elsewhere.

Before the transport revolution, which was contemporaneous with the industrial revolution, sailing ships had made sufficient progress that all the world's shores were potentially part of the same market at the global scale. Products manufactured in England were for sale everywhere and could travel more easily now that mechanization had lowered their cost: local artisanal articles were eliminated. It was a great temptation for importers to copy the British example and in their turn to launch into manufacturing. European societies possessed all the artisans who could easily be converted to industrial work, and businessmen ready to invest their savings in firms where high profits could be made. Looms, spindles, railways and steamboats were thus adopted very quickly by the countries of north-west Europe and by the United States – in this case, the backlog behind England was often only a decade.

The diffusion of progress was not confined to Europe and North America: it rapidly reached other continents. Countries which until then were practically empty were opened up to white settlement (at the time the term 'new countries' was employed): the demographic explosion was so rapid in Europe that manufacturing was insufficient to absorb population increase. In these new lands where the settlers had a sound technical background, the application of new machines was as rapid as in Europe and affected transport, agriculture, livestock-rearing and mineral extraction. Lacking local markets, industrialization was not possible. Regional organization was rapidly marked by the development of specialized farming areas, punctuated by small centres indispensable for the commercialization of production and the supply of services necessary for the new farmers. Concentrations of population occurred in the mining zones. But major urban centres were rare: spatial organization was dominated by the ports which ensured the link with the import areas in Europe. Where development penetrated inland, a few major nodes also emerged on the transport networks. Australia, New Zealand, Argentina and the Canadian West best exemplify these processes.

Where the traditional societies were more densely present, European colonization sometimes created conditions reminiscent of the new countries: farmers arriving from the mother country bought a proportion of the lands belonging to the indigenous population or settled on land from which they had been expelled following revolts. Thus in the cases of Algeria, Morocco, Tunisia, South Africa, Rhodesia (present-day Zimbabwe) and Kenya, a division into two spatial sectors was observed: one open to white settlement, which evolved along the lines of the new countries, and the other where the indigenous population was contained, which was slow to adapt to new methods.

It was in South Africa that this form of spatial organization was best seen: the white farmer country of Cape Province, Orange Free State and Transvaal, as well as the narrow coastal fringe of plantations in Natal, were specialized agricultural zones dominated by ports (Capetown, Port Elizabeth, East London, Durban) or by transport nodes, which also attracted political functions (Pretoria and Bloemfontein). A mineral and industrial region was born in the Witwatersrand, around Johannesburg, from the exploitation of gold seams. The zones where the Bantu predominated extended in a half-circle around the white areas, to the north and east. From Swaziland to Lesotho and to the Ciskei via the upper part of Natal and Eastern Transvaal and by Bechuanaland, they follow the uplands from one part to another of the great Drakensberg escarpment. They retain a traditional African rural aspect even when the inhabitants leave to earn their living in the mines. Infrastructures and an urban network were absent, whereas they were highly structured throughout the colonized space. The contrasts made concrete by apartheid after 1948, were thus laid down from the beginning of the century, by the strength of the contrasts between the zones occupied by colonists and those where the continuity of Bantu settlement had not been destroyed.

The diffusion of the new techniques, which operated without problems in countries populated by Europeans, elsewhere confronted considerable obstacles. They came sometimes from the local powers that refused to open their countries to international trade and to imports of manufactured goods from Great Britain and Europe. In a few years, armed diplomacy generally prevailed over these forms of resistance: China and Japan were thus constrained to renounce isolationism in the 1940s and 1950s.

But the suppression of measures forbidding imports was not sufficient for growth to take off. Some countries looked favourably on the modernization of their economy: it was indispensable in order to resist European ambitions. Turkey, Egypt and Japan attracted experts to copy European technology. Japan was the only country to succeed in this way. Although its civilization was profoundly different from those of the West, it possessed certain characteristics in common with them. Its cohesion was strong; national sentiment created a deep solidarity between all classes; the level of education of the whole population was good; an indigenous commercial tradition existed which eased the development of enterprises copied from the European model, and had artisans to supply the future skilled workers.

Elsewhere, the grafting of techniques did not succeed; education was not sufficiently widespread to provide for commercial training and, in the absence of scientific and technical training, it did not produce engineers capable of operating the new manufactured goods. The mass of the population refused to bow before the disciplines imposed by the use of machinery. The training of skilled workers was almost impossible: locally there were none of the types of artisan trade available to facilitate the acquisition of skills. To import European workers was expensive and insufficient: the indigenous population was incapable of taking advantage of them to imitate and replace them. The very wide gap between the elite, which drew its wealth from public service and landholding, and the mass of the population limited market size and discouraged initiatives.

The Europeans were proud of the gulf between their economies and those of the rest of the world. By virtue of this they extended their domination to virtually all of Africa and a large part of Southern Asia, and progressively infiltrated the Ottoman and Chinese Empires.

Spatial organization outside modernized nations

In the nineteenth century, the organization of space, which became established outside Europe and the new countries, signified a dualism: a dualism between the mass of the indigenous population and the class that organized and governed it (this consisted of autochthonous indigenous elites or elites imposed by earlier conquests, or an administration put in place by the colonial power, or a mix of these different categories); a dualism in the organization of space followed since modern features were imposed in zones open to white settlement or those where expatriates controlled plantations or mineral exploitation, while in the rest of the space life continued without any apparent change.

The organization of space by modern networks of transport and communications and its opening up by service centres required by a modern economy was thus confined to the ports, which carried out contacts with the exterior, and to zones producing for the world market. Elsewhere, penetration was more restricted: the colonial or indigenous commercial companies which prospered by selling imported manufactured goods did not always need heavy transport systems, and relied on traditional routes and centres which they scarcely changed.

The Dutch transformed the East Indies islands. Over most of the archipelago, their presence was modest and their impact slight. Their colonial effort only commenced in Sumatra at the beginning of the twentieth century. By contrast, in Java (figure 11.3) it was early and deep, but varied in its nature. The agrarian landscapes and social structures of the eastern plains and basins (around Surabaya) and the centre of the island (whether the interior basin of Surakarta or the open plain to the south of Jogjakarta) were scarcely changed. The administration limited itself in these regions to encouraging the farmers to indulge in commercial crops, sugar cane in the east around Surabaya, coffee and tobacco in the centre, where very high population densities limit development. In the west around Jakarta, colonization encountered different conditions: plantations of tea, then of rubber, extended widely at medium altitude over the volcanic slopes. They progressed from there to southern Sumatra.

Dualism is expressed here by the contrasts in the methods of colonial development in the east and the centre on the one hand and the west on the other. But as the whole island furnished export products the transport infrastructure is evenly distributed. Sumatra offers a more classic model, since the areas of traditional population, dense in the mountains and sparse in the eastern plains, did not receive railways and lack roads. Modern infrastructure did not extend beyond the plantation zones.

The traditional Javanese world was dominated by the capitals of the ancient Hindu kingdoms, Jogjakarta, Surakarta, etc. In the colonial era, these towns were still administrative centres and local markets, but growth was above all to the advantage of the ports opening onto the world market, Surabaya to the east and Jakarta (formerly Batavia, a name recalling its Dutch foundation) to the west. Bandung owed its prosperity to its climate, modified by altitude, which suited the colonials.

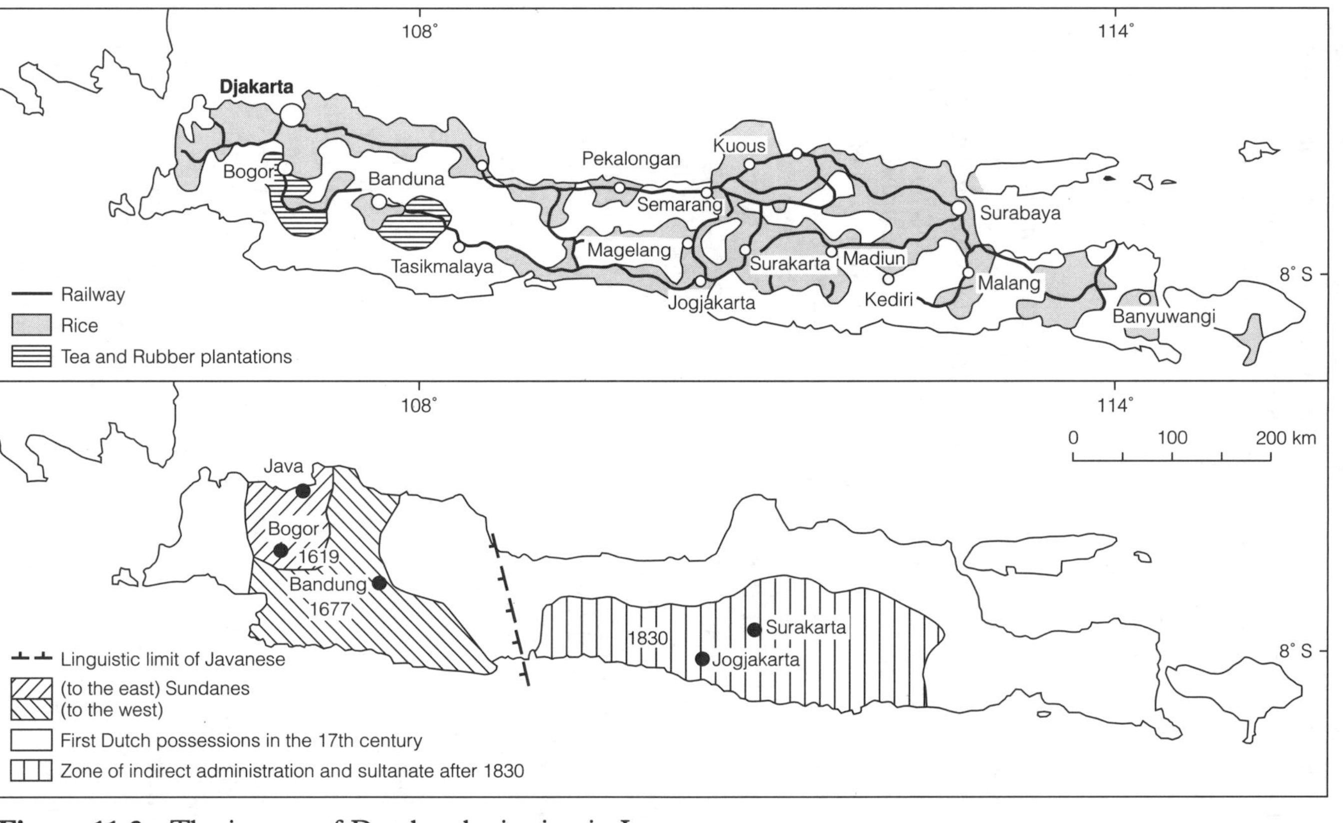

Figure 11.3 The impact of Dutch colonization in Java
In the east and centre, colonization imposed commercial crops into countryside carefully organized by the sultanates who for long remained independent. In the west, colonization, much earlier, did not find the same kind of advanced agriculture. It set about intensifying rice cultivation by irrigation and developed plantations in the hill and mountain zones.

Social effects and reactions resulting from the Europeanization of the world

The social changes flowing from the apparently modest forms of penetration were much more profound than a superficial examination might suggest. Imported goods ruined a proportion of traditional artisan activity to the extent that employment linked to primary resources became higher than it was prior to the explosion of progress. Social stratification tended to shrink, accentuating again the dualism between the dominant classes and the mass of the population.

Europe and the United States were so powerful that everywhere in the world populations were fascinated by their culture: if there were some cases, in China especially, where the power opposed the penetration of the philosophies, religions and techniques from industrialized centres, the dominant attitude was different. A large part of the elite sought to westernize: they sent their children to institutions run by priests, pastors or laypeople coming from Europe or the United States, they tried to adopt modern science. It was difficult to convert to Christianity wherever highly structured religions existed but, throughout the nineteenth and the beginning of the twentieth centuries, many indigenous people lived a double allegiance to their traditional religion and to modern forms of progress, apparently without problems.

In many cases, fascination with the West, however, challenged traditional values. The morality in which one had been raised was broken and replaced with others. In South America, it was marked by the conversion of the elites to European secular rationalism: a part of the Creole aristocracy adopted the cult of progress passionately. In Brazil, where the phenomenon was more marked than elsewhere, positivism became almost a state religion, at least for the ruling elites, whereas the rationalist spiritualism of Cartec achieved a lively success in the middle classes, highly influenced by spiritist forms from Africa, but were content to purify it. Such conversions led to new developments of space and to new definitions of the sacred and the profane.

Sometimes, the culture shock of the West was even stronger: the elite was not alone in being fascinated by the power of modern techniques. This explains the number of conversions in societies where there was no religious tradition to oppose the diffusion of Christianity. But heresies appeared rapidly, strange syncretisms saw the light of day. In the case of the Caodaïstes* of South Vietnam, elements borrowed from Buddhism were mixed with Confucian reverence for written knowledge, the cult of saints inspired by Christianity and the strange canonization of western scientists. The new forms of faith

* A religion founded in 1926 by Ngô Van Chiên.

often assumed more uncouth characteristics. They did not favour the emergence of westernized elites, but they explain the vigour of certain social movements – the revolt of the Taipings, for example, in Yangtze China from 1846: its instigator gained his inspiration from what he learned from missionaries, which explains the brutality with which he broke the social, philosophical, religious and spatial order underpinning the Chinese Empire since the Han dynasty.

These 'cargo' cults, as anthropologists later termed them, prove that the vision of the world held by societies opened up to European trade already differed greatly from the traditional model. A secularization, at least in part, of the environment resulted, and permitted the first stages of modernization. These movements generally appear as shoots of Christianity grafted onto a local base, but it can happen that other religions are transformed – the *mouride maraboutism** in Senegal provides an excellent illustration.

11.2 The Apogee of the Economic Nation and the Centre–Periphery Model: the Example of the United States

The new conditions of growth

After 1870, new forms of spatial organization appeared in countries where the take-off occurred one or two generations late, in Germany and in the United States (where the Civil War precipitated industrial growth). Great Britain remained loyal for a long time to the open model that had given it prosperity and which free trade had supported; the advances that had been achieved in overseas markets, the solidity of the commercial networks that it had established over wide areas and the vertiginous expansion of the empire throughout the nineteenth century meant that until the interwar period Britain need not question its economic orientation and the forms of regional organization that flowed from it.

The conditions had, however, changed significantly. Productivity gains became apparent in agriculture where mechanization finally began and where fertilizers permitted intensification. The range of industrial products widened: to the traditional consumer items, cloth, furniture, domestic utensils, a wide range of materials and transport vehicles (bicycles and the motor car for example) were added, as were tools for agriculture and machinery for industries. Chemical industries diversified the raw materials for industry and created new markets – synthetic medicines in particular. The production chains became longer:

* A *marabout* in the Muslim religion refers to a holy man or a shrine.

an increasing number of firms specialized in manufacture which was incorporated by other factories into articles for final consumption or as capital goods. The significance of intermediate production increased rapidly. Since the contacts between the different stages within a production chain are incessant, it was impossible, given the means of communication and travel at the time, for them to be located too far from each other. The internationalization of markets was apparent in the field of raw materials as with capital and consumer goods, but the market for intermediate goods remained national: this strengthened the dynamism of the nations that were the first to modernize.

The diminishing cost of transport accelerated world-wide due to the railways. The regularities of services that it permitted facilitated the distribution of products over widespread markets. Migration became easy. The artisans ruined by new industrial products left for the towns in search of work. Landless peasants participated in the movement as the use of machinery reduced the range of tasks available to them on farms and made their employment more and more irregular and uncertain.

The organization of national spaces

With lower transport costs, the choices open to industrialists to locate their factories became diversified: activities with heavy energy needs and those that involved a major reduction in weight in the raw materials used, remained tied to zones of energy production or raw materials — the coalfields remained very attractive. In the case of products where the majority of their transport and communication costs were related to marketing and distribution, central locations became advantageous. To the extent that a general increase in living standards permitted by increased productivity created new internal outlets, exports ceased to be the sole motor of growth: growth depended on increasing the national clientele, which opened up to include more varied consumption. For this type of evolution to be rapid it was sufficient to ensure that the vast majority of income engendered by increased demand should be spent within the country: this encouraged renewed waves of purchasing and an increased diversification of production.

Development thus accelerated when national economies were not too open and benefited from the effects of increased incomes. The era of free trade ended. National markets became indispensable for the appearance of continuous growth. Within national markets specializations strengthened due to competition and central regions saw their activities diversify as external economies multiplied.

As the population employed in agriculture was still large, the general mesh of spatial organization remained very marked by the distribution system necessary for rural areas. The structural frameworks established in the traditional world remained apparent, but the weight of the urban population increased and the metropolises which structured these economic spaces became totally open, but protected from very keen external competition; they saw their functions and populations take off. Industrial production remained grouped in relatively compact specialized regions, but they became more diversified in nature: to the zones which enjoyed the best accessibility to the entire market were added nodes of heavy industry located close to energy sources, areas where light industries diffused into the rural world and urban centres, which grouped activities, highly dependent on information.

The example of the United States

The United States offered, in the first half of the present century, the most complete example of regional organization of the type described above. The pattern of locations is clearer than elsewhere in the sense that specialization was established on virgin territory without needing to take into account the heritage of periods when the economy was organized according to more traditional principles. In the United States there were no old provinces where solidarities had taken root to deflect the new currents of relationships. Industrialization was early, but occurred after the railways came into existence, so that there were never really any 'black countries' on coal or iron ore fields. The distribution of energy resources, water power, coal then petroleum, was such that no region possessed a monopoly that gave it a decisive advantage; none of them lacked the basis indispensable for modern activities either.

After 1790, the American government had installed an effective protection of industry that permitted multiplier effects and linkages. All the conditions were thus combined for the new dynamic of location to find its maximum expression. The economic structuring was clear everywhere: production came from specialized areas and the networks dominated by powerful metropolises, organized the whole system of relationships.

It was in agriculture and livestock farming that the results were most spectacular (see chapter 5, figure 5.2). The South, from the Atlantic coast to Texas, was almost entirely devoted to cotton (the Cotton Belt), with, on its periphery, more restricted zones devoted in the south to rice or sugar cane (Louisiana and coastal Texas), citrus fruit (Florida), and in the north to tobacco (the tobacco belts of North

Carolina and Virginia on the coast, and of Kentucky in the interior). Cereal-growing was the major speciality of the interior plains. Maize made Ohio and Iowa, the stronghold of the Corn Belt, which used it to fatten pigs and the underweight cattle from the west. Wheat came from the dry prairies: the Wheat Belts (there are two, separated by the dunes of Nebraska) extend from approximately the 96° west line of longitude to 100° to 102° west – the limit has continuously fluctuated according to the dry and humid periods. In the south winter wheat predominated. In the north, agriculture was initiated later: it was not possible until the introduction of Russian spring wheat in the middle of the 1880s. A tongue of the Wheat Belt extends on to the Columbia plateaux in Washington State.

Cattle raising is split between the Dairy Belt, which extends from New England to Minnesota, in the regions of cool summers in the north, and the ranching zones of the west, devoted to sheep and to the raising of calves for later fattening in the Corn Belt. The Atlantic and Pacific seaboards are devoted in part to the production of vegetables, and in the case of California, fruits and vines.

Specialization equally took other forms: mining in the iron ore zone of Lake Superior or in the non-ferrous deposits of the Rockies, petroleum in the south and west, tourism where climate and environment attracted visitors. It was difficult in Europe to imagine that certain regions could live from tourism alone, emptying each year at the end of the season. However, this was the case very early in the United States: on the cool summer coasts of New England, the massifs of the White Mountains, the Catskills and Adirondacks (in New York State) as well as Southern Florida, Southern California (originally), the Rockies front range, around Denver and Yellowstone Park.

One ensemble, however, can be detached from the list of specialization: the Industrial Belt, which extends from the Atlantic Coast, between Boston and Baltimore to the Mississippi (and a little beyond), between Minneapolis and St Louis (figure 11.4). To the agricultural orientation of this ensemble (which is superimposed on the richest sectors of the Dairy and Corn Belts) were added the essential manufacturing activities and service centres of the country. The Industrial Belt coincided with the highest zone of population potential of the United States (figure 5.3); California, remote but with a rapidly growing population, created a core with a lower potential but sufficient also to attract manufacturing.

When the Industrial Revolution took place, the demographic dominance of the north-east was strong since this is where immigrants had always arrived and tended to accumulate. This explains the location of the Industrial Belt; it is located at the heart of the national market. Further west colonization took place after the arrival of the railway, so

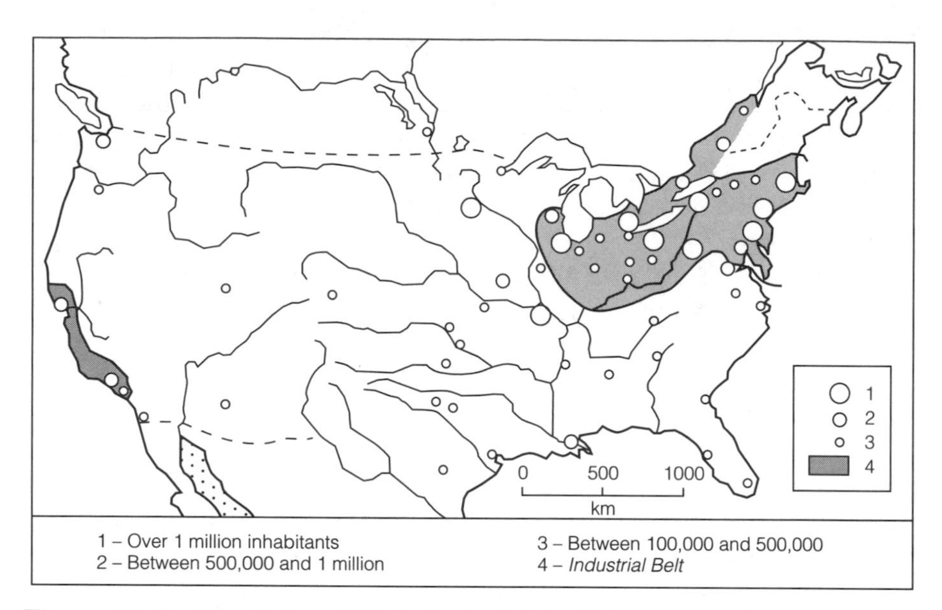

Figure 11.4 The industrial belt in the United States towards 1930
The essential manufacturing activity and the great majority of service activity related to firms, were concentrated in the industrial belt, from the North East to Mississippi and its Californian annexe.
Source: P. Claval, *La Conquête de l'espace américain*, Paris: Flammarion, 1989, p. 159.

that artisan activities had not flourished, which had supplied skilled labour in the east and the entrepreneurial traditions necessary for the great economic mutation. Development occurred when it was already easy to bring in whatever was lacking. As a result, the creation of industries was not encouraged and certainly contributed to the difficulties encountered in establishing manufacturing industries, but the eccentric location weighed even more. The contrast related to the progression of the pioneer frontier was also apparent in the structure of polarized regions: born to gather products for export and to deliver the tools, equipment, people and capital necessary for development, their capitals generally had spheres of influence more extensively oriented to the west than the east.

In the states of the Industrial Belt like Ohio, the regional capitals (figure 11.5) are generally close to each other, since there are manufacturing regions with high densities that they organize, and markets of national or international extent that they structure. On the Atlantic coast, which maintained the essentials of international relations of a country still very much oriented towards Europe, the cities almost touched each other: they formed a practically continuous string – Megalopolis, as Jean Gottman termed it. It was a new phenomenon;

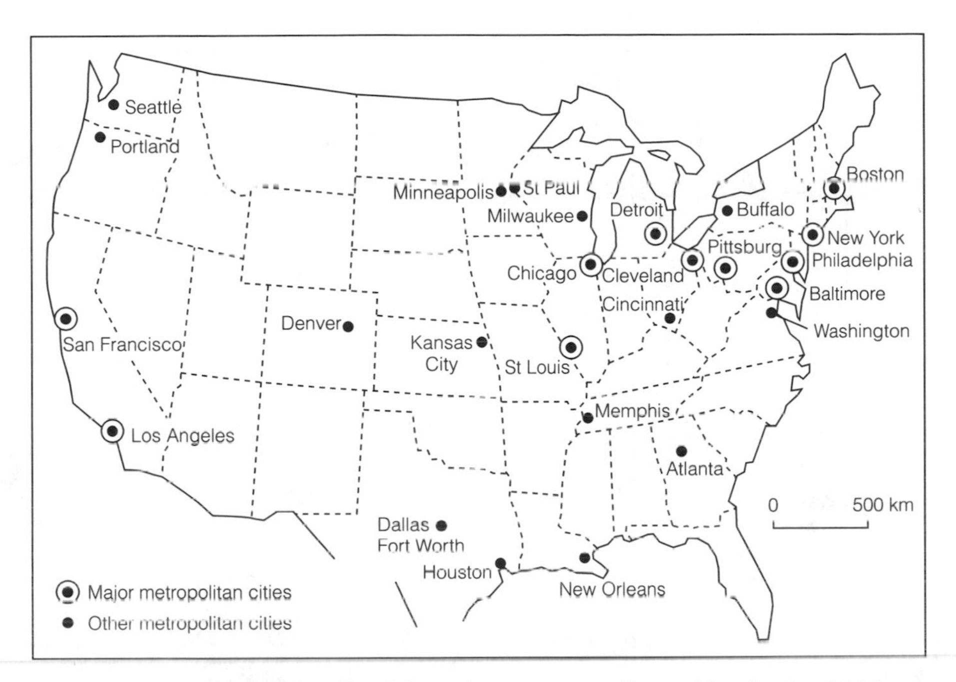

Figure 11.5 The network of American metropolitan cities in the 1930s

located there were major centres concentrating strongly on service activities, but which did not have large direct hinterlands. Their role extended to the scale of the United States, of the entire continent and even the world.

The impact of the new model in Europe on the rest of the world

In the old countries of Europe, the establishment of new territorial organizations gave birth to more complex distributions since those inherited from the pre-industrial era or created during the first phase of modernization weighed more heavily. But the characteristics, so clearly seen in the United States, could also be recognized there: agricultural or mining specialization was almost total in the periphery and the concentration of manufacturing activities was accentuated in the central areas – in France the part of the country to the north of the line Lyon – Le Havre.

The British example is especially telling (figure 11.6). Until the First World War the economic system remained very open: Britain was loyal to the plan of the first phase of the industrial revolution which translated into a peripheral distribution of the major manufacturing sites. The First World War brutally upset the situation. Britain resolved not to devalue the pound and succeeded, thanks to the energy

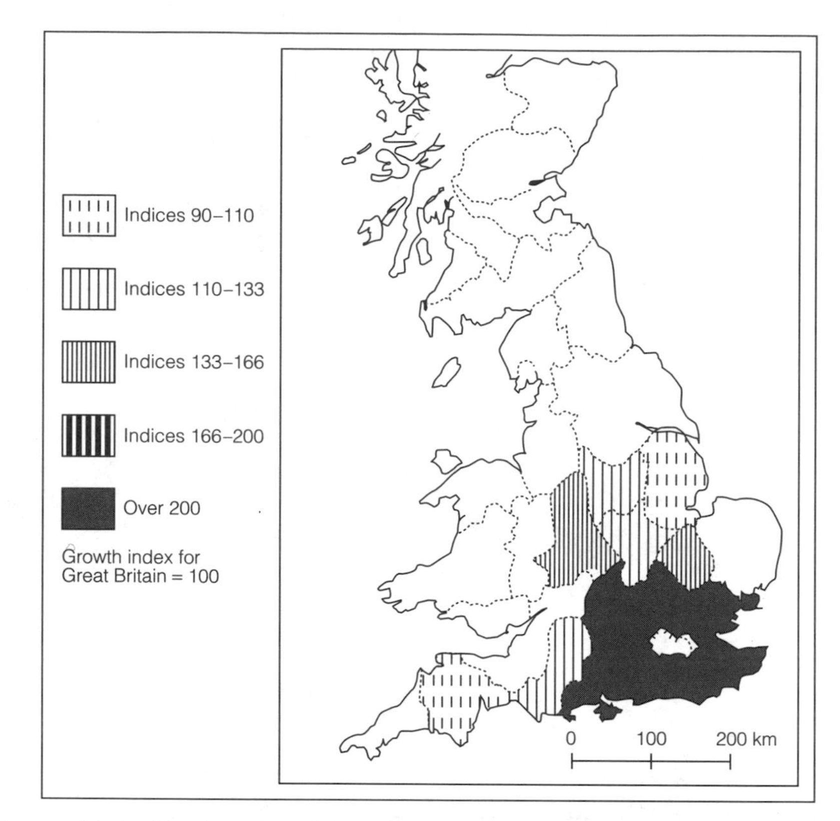

Figure 11.6 Regions in crisis and dynamic regions in England as evidenced by demographic change from 1921 to 1939

Growth was concentrated in the South East and the Midlands, while the peripheral industrial regions were all experiencing crisis.

of Winston Churchill, in re-establishing her role as before the conflict: this policy was necessary to maintain the international role of London as a world financial market and to support the sector of the British economy that lived off international services and the imperial function. But the lack of devaluation caused a lack of competitiveness in the majority of the manufacturing industries born in the nineteenth century as it was impossible to revoke the substantial wage increases granted during the war: the exports of coal and cotton fabrics were impossible and those of steel and ships suffered increasing difficulties. Unemployment appeared after 1920 in the industrial zones of Scotland, the North of England and Wales: it has remained ever since, with the exception of a brief interruption during the Second World War.

The British economy experienced at the same time a vigorous growth in certain manufactures: these, as elsewhere in the developed world, were linked to an enlargement of the internal market stimulated by increased incomes. These were durable consumer goods, which benefited from the new growth. Their manufacture was concentrated in the Midlands and the London region: car production developed in Birmingham, Coventry, Oxford and at Dagenham in the suburbs of London. Within an England in general decline from now on was opposed an expanding zone that coincided, as everywhere, with the highest levels of population potential. In Europe, a certain degree of specialization was defined between the nations. To the extent that the logic of protectionism was never pushed to its limits, for certain manufactures the continent was already assimilated as a single space within which the central areas were favoured. It was the origin of the strengthening of the axis which runs across the heart of the continent from the London basin to northern Italy.

The contrast between the centre and periphery, remarkable at the national scale, detectable in Europe at a transnational scale, also structured the world space: the industrial economies drew on the whole world to procure raw materials and the agricultural food supplies that they could not produce. Protectionist measures that they surrounded themselves with were sufficient to condemn to failure every whim to launch manufacturing activity in peripheral countries. It was difficult, as well, to assure the diffusion of new methods of production towards areas lacking the essential know-how for manufacturing and the attitudes without which firms go bankrupt.

Regional space: new problems, new structures, new representations

In the course of the second phase of the industrial revolution, the problems of pollution seemed less dramatic than in the previous phase, since industrial areas tended to spread out it was easier to transport energy. The widespread use of electricity permitted a separation of the zones where factories located from those affected by ash and gaseous effluent from the fuels used by steam turbines in power stations. Techniques in urban planning had made considerable progress. Less dense districts were now planned where microbian pollution declined. Water and sewerage systems extended beyond the parts where the impact of urban ecosystems was felt. Disequilibria existed, as we know, but they were limited to narrow zones in the largest cities. Measures were taken to limit the toxic dust and gas waste from industrial and domestic sources: in New York, for example, the less-polluting anthracite was the only authorized coal. The modernization of agriculture had not yet led to a sufficiently massive use of fertilizers that the natural equilibrium of rural regions seemed menaced. Public opinion, even the informed, was not conscious of the ecological strains that were nevertheless imposed in many regions.

It was in the course of this second phase of regional construction that the rationalization and secularization of the world progressed furthest. Almost all space was still devoted to productive activity: the value of places was based on an estimate of their economic viability. The abundance of mineral resources, soil fertility, the advantages that climates gave for particular types of agricultural production and market accessibility, were the essential arguments invoked by geographers who considered the regional structure of fully commercialized economies.

Increased mobility had facilitated population movements, the development of industrial regions and cities, but, with a few exceptions and only for very marginal areas, without ending up with the emptying of extensive areas. Daily commuting only extended where the density and structure of population made public transport systems worthwhile. In their essential features, the daily life of industrialized areas was still set, between the world wars, in the framework of the cells inherited from the traditional rural world. Urban growth was facilitated by the extension of movements to work or for purchasing permitted by the railway and tram lines, but low-density suburban developments were excluded since they did not generate sufficient traffic to amortize the necessary infrastructure. Growth therefore proceeded in the shape of fingers of a glove, with urban antennae separated by zones where countryside persisted.

The productive space is only valued as a function of its resources and the opportunities it opens up for contacts. In perimeters devoted to residential use or regions devoted to recreation, the evolution takes into account more subjective dimensions. This was readily observable by many signs. Churches and chapels proliferated in the suburbs and in industrial agglomerations and the local powers took care to underline the majesty it must surround itself with by choosing for its prefectures, town halls and police stations, a style and dimension that contrasted with buildings of purely utilitarian purpose. The prestige attached to culture was seen by the monumental character of theatres, opera houses, museums or universities. The cult of nature was evident in the care taken in the design of parks and public gardens. Harmonious shapes and colours were desired there. The appearance of natural parks and reserves where the vegetation had retained its natural character demonstrated that the effects of new forms of awareness extended beyond urban zones. The aesthetic spirit was also apparent in tourist areas favoured by the rising living standards: their clientele progressively widened to include the middle and working classes, which explains their continuing extension and the less elitist character of their scenery.

In the West, the dogma of progress was never challenged, even if the means of achieving it were debated as in the case of socialist movements. Enterprises owed their prestige to the success that they experienced and tried to affirm it by the construction of headquarters or production plants displaying their new power: the skyline of American cities' office towers progressively obliterated the church towers and courthouse domes or administrative buildings. The dominant ideologies gave pride of place to Utopia: the places where they flourished, those where planning already evoked a glorious tomorrow, carried real prestige. At the same time, nationalist fervour explained the lay cult of martyrs of the country. After the First World War this led in even the most isolated communities to the erection of war memorials; not without emotion one discovers on a plaque, in a remote hamlet in South Island, New Zealand, the names of those who fell to defend a certain idea of democracy or liberty in Flanders or on the Somme during the First World War.

11.3 The Emergence of Underdevelopment: Parallel Blockages and Avenues

After the 1880s and between the two world wars, the construction of dynamic national economies slowed down the diffusion of progress: the economic growth of industrialized countries no longer required the ceaseless conquering of new external markets to absorb their output. It depended on growth in national incomes, on the endless and more diversified demand for goods that it provoked and on the resultant stimulus to production. The enlargement of internal markets born of increased wealth meant that it was easier to benefit from economies of scale whilst the complexity of production circuits multiplied externalities. Directors of companies in this situation had no interest in locating outside regions that were already serviced.

The difficulty encountered by the diffusion of industrial know-how followed the same logic: even when wages were minimal, labour costs were high outside countries which had already achieved development at the end of the nineteenth century, since even very elementary qualifications were absent and the means of teaching them were lacking.

During the first half of the twentieth century, the geography of development seemed to be frozen. The geographical extension of industrialized space almost stopped – only the countries of the New World of the previous epoch saw their industrial activities progress: the British dominions, by their occupational structure, resembled more and more their mother country, even though the size of their internal markets limited part of their manufacturing to assembly work. Where European settlement had only progressed by superimposition on, or displacement of, already dense indigenous populations, the markets open to consumer durables were more limited, but the higher revenues of the colonists benefited the repairing and maintenance artisans and the firms making transport infrastructure. In Latin America, industry, stimulated temporarily by the First World War, came to a halt: even in countries the size of Mexico, Brazil or Argentina, the internal demand was too limited, and the difficulty of internal communications sometimes prevented the maximum advantage to be taken of all the possibilities, as in Brazil.

Variations on the theme of social and spatial dualism, which since the middle of the nineteenth century characterized countries where industry had not taken hold, became fully asserted. The equipping of plantation areas or mining fields improved and the services offered by ports serving as bridgeheads for modernization diversified. Elsewhere, the structuring of space by the ruling society remained slight: the development of indirect administration in colonies allowed traditional structures to persist almost intact.

The rise of resistance to westernization

Reactions in the face of modernization multiplied. Indigenous elites were always fascinated by European technology but the dangers born of the loss of cultural identity were better understood. The number of indigenous people educated in secondary schools or in western universities increased; a proportion of the elite became comfortable with both cultures. Within them, movements reinterpreting traditional values developed with double-edged objectives: to catch up with the West and its technology, and to base life on modernized traditions. Gandhi was perfectly well aware of the superiority of European technology but he militated in favour of a return to traditional artisan products to reduce the economic dependence of his country and to avoid the elites sacrificing the masses without whose support the struggle for independence would have been impossible. He also tried to limit the disruption that an uncontrolled modernization would impose on spatial organization in India.

It is in this context that the formation of the first fundamentalist movements were to be seen. In Egypt, the Muslim Brothers preached a return to permanent values of Islam. They feared that by indiscriminate importing of western culture, Muslim society would be lost. From then on cultural borrowing was assiduously supervised. The behaviour and morals of the dominant world powers became the object of pitiless criticism.

The major new feature of the second phase of world industrialization resided, however, in the appearance of socialist regimes. The concentration of industrial labour in 'black countries' and in cities at the beginning of the nineteenth century took place in inhuman conditions that very soon were condemned and very soon fed the strikes and uprisings that punctuated the history of the nineteenth century. The thinkers who built dreams of Utopias believed in progress; their objective from this point of view did not differ from the Liberals but they proposed other means of achieving it. The injustices springing from working conditions and income distribution could not be avoided except by renouncing market forces: the decisions relating to production and distribution of products were too serious to be left to individual initiative. They should be vested in the government, alone capable of ensuring the triumph of the general interest, in the face of the self-interest, of all the groups participating in production.

The October Revolution in Russia transformed the socialist programme from dream to reality: from then on there was a country the development of which would be organized around a purified social and economic programme. The cultural and national reactions that the preponderance of industrial nations solicited outside Europe were from now on expressed in a discourse that respected the ideas of progress, but condemned all the institutions that had permitted seizure by the West to take place. Anti-colonial movements were stimulated by this sentiment. They used a language borrowed from modernity and leant on the messianism of all the religions of progress. The new revolutionary programmes were all the more credible in that they had recourse to the primacy of the West in all domains. Quoted extracts as a source of inspiration for independence movements, from Christian messianism to Communism is clear in the case of Sun Yat Sen: the Guomindang on which he leant when he founded the Republic in 1911 still carried the mark of the Protestant teaching that he had received. In his struggle to reconquer power, he discovered the force of Marxist thought and the help that the USSR could give him.

The socialist system and the organization of space

The founding principles of socialism kept silent on the subject of the spatial organization of the future; Marx discouraged speculation on this theme by denouncing Utopian trains of thought which were the most direct competitors for the ideology he proposed. One thing only is certain: it was necessary to condemn the market. The strategy of the transition to socialism was, moreover, not envisaged in a spatial context. It was expressed in terms of the global economy: the success of capitalism derived from the multiplication of productive capacities that flowed from the use of concentrated forms of energy and the use of machinery. To fight capitalist societies on their own ground, it was sufficient to accelerate the formation and accumulation of capital and the arrival of the age of abundance where society could really be restructured.

This analysis at first sight seemed convincing, but it ignored the spatial dimension of the processes that lead to growth. It was not sufficient, in order to make progress, to multiply productive capacities. Simultaneously it is necessary to organize material supplies and to assure the distribution of production: the industrial revolution was based on both the mobilization of energy sources and new technologies, and on the improvement of transparency resulting from the transport revolution and progress in all forms of communication. Socialism left its mark on this aspect of communication. By condemning the market, it necessitated central planning, which extended the circuits of information, harmed their quality and considerably encumbered the process of arriving at decisions.

To make the system function, the regime quickly had to invent palliatives: by assigning certain agricultural units to supply such and such factory or the inhabitants of such and such a town, an economy was built with long information circuits, but it was deprived of the adjustment permitting the confrontation of supply and demand based on a large market; the stimulus of competition was suppressed. Priority was also given to the gigantic size of industrial plants: this was expected to give an easier control of decisions taken by the party, a reduction in the number of production units for which the planners had to establish production plans and to direct exchanges, and the internalization of the manufacture of components and sub-ensembles. The result was that the large enterprises were often no more than a collection of medium-sized workshops: the management controlled the quality of what they produced only by calling in external enterprises, but this immediately denied the economies of scale possible and healthy competition between subcontractors.

With no land costs, there was no information available on the external economies that might be expected from one location or another; overall growth was slowed down since plants took longer to start up than if they had been correctly sited. They were expensive and were more often affected by breakdowns or shortages of supplies than if they had had maintenance services and subcontractors close by.

By refusing to make large investments in transport and communications, the Soviet planners attempted to choose a short-cut in the search for progress. They in fact neglected everything that permitted the inclusion of space in networks established where circulation was easy. The towns lost their role as trade centres and as decision-making places, in favour of the capital city towards which all flows converged. They contrived to supply education and health services and certain commercial facilities, but this was all that remained of their traditional functions. This induced the decline or lifelessness of the lower levels of the urban hierarchy. In the centres benefiting from industrial implantations, population growth by contrast was rapid, which misled western observers. Socialist towns owed their dynamism to the sometimes gigantic factories that they hosted: they did not have strictly urban functions. The capital owed to the presence of central planning services the almost total monopoly of responsibility for the structuring of space. In this sense, it remained a real city, which explains its attractiveness for all the population. However, the capital was in a situation of generally collapsing under the responsibilities placed on it: its circuits of information were permanently saturated and delays accumulated.

Centralization made it impossible to take into account in location decisions all the indirect effects they could have on the natural environment and all that the social environment could also contribute: positive externalities were not exploited and the negatives (in the form of pollution) multiplied without the local population, the most directly concerned, having the slightest means of action.

The geography that began to be drawn in the USSR in the 1930s was then very different from the countries with liberal economics. For the Soviet theoreticians, the superiority of their model was supported by a better consideration of the general interest. Analysis of the methods of functioning showed that its effects were exactly the opposite, but no one wished to admit it. The isolation in which the USSR existed prevented a critical judgement to be made of the real methods of organization. Initially, the priority given to investment had tangible effects: the production of energy, minerals, steel or wheat increased substantially. Difficulties only appeared with the diversification of production after the Second World War.

Conclusion

The world's industrialization resulted from the improvement of transport and communications, but its first result was to increase the contrasts between levels of development and the modes of spatial organization. The traditional regional structures, although progressively weakened, remained in place in the countries that were not transformed by industrialization. Around the ports and zones oriented towards exporting, one saw, however, the appearance of urban-based regions, but they only enclosed a small proportion of space.

In the industrialized world, it was around regional capitals that national spaces were ordered, but the dimension and role of the areas that they controlled varied according to their position. In the peripheral sectors, they marketed agricultural products and raw materials or heavy semi-finished products that comprised their resources. In the central sectors, the areas that they controlled were penetrated by manufacturing industries. They also lived by services with distant or extraterritorial markets. The support offered by nearby areas, which all capitals need, practically disappeared – they served only to supply the essential water supply, to weekend leisure areas and sewage farms, for used water. From then on, cities could proliferate without their closeness to each other handicapping them.

By facilitating transport and by substantially reducing the proportion of the population employed in the production of primary resources, the industrial revolution gave a spectacular boost to interrelationships and considerably strengthened the role of networks, as evidenced by rapid urbanization, the increased hierarchical structuring of towns and the new roles of regional capitals. But this opening up was also translated into the formation of large homogeneous areas of specialized activities and by their articulation around cores constituted by the central areas of economic zones.

Two factors still limited the evolution: the high proportion of the population which remained employed in the primary sector, and the imperfections in the communications systems.

(1) The first element explains why regional organization generally was framed in the continuation of the pre-industrial situation (except in the central cores of nations which had generally been reorganized, and in the industrial zones), since it was still advantageous to draw from the agricultural workforce for labour needs.

(2) The second reinforced the role of nations, since it was impossible to organize at a larger scale certain aspects of direct contact: the production of intermediate goods and services escaped internationalization.

The relative slowness of communication was especially inconvenient for business where, very often, direct contacts are essential to secure markets. This enforced a fragmentation of the linkages at the world scale between producers and consumers, and to the multiplication of stockpiles over the length of circuits. This explains the role played by ports and their specialized groups of shipowners, dockers, importers and exporters throughout the first phase of industrialization. If a world economy already existed then, it was due to powerful organizations which, on the North Sea and the European and North American Atlantic coasts had the ability to weave effective networks across the planet.

Internationalization of economic life had made considerable progress, but the networks that made this possible were managed by specialists who lived at the margin of nations, in their great ports. This doubtless is why the enlargement of circuits did not yet upset the nature of feelings of belonging. Those attached to the regional scale of life did not disappear, but became integrated at an ever strengthening national scale of identity.

The enlargement of interactions and the easier transmission of knowledge, ideologies and conceptions of existence that they engendered, however, had already transformed the world's cultural geography. In all the continents the Europeanization of the elites and the reactions it caused were to be observed. The belief in progress meant that the growth of internationalization was universally accepted.

This evolution was heavy with tensions: this was fully realized in the 1930s when the abandonment of the gold standard produced economic global disintegration. In the aftermath of the Second World War, the urgent need for the reconstruction of Europe, Japan and China, and the recognition at last of the need for the development of the Third World gave a dramatic dimension to spatial organization.

Further Reading

Clapham, J. H., *The Economic Development of France and Germany, 1815–1914*, Cambridge: Cambridge University Press, 4th edn, 1963, 420 pp.

Clout, H. D. (ed.), *Themes in the Historical Geography of France*, ch. 12, 'Industrial Development in the Eighteenth and Nineteenth Centuries', and ch. 13, 'Urban Growth, 1500–1900', pp. 447–540, London: Academic Press, 1977, 594 pp.

Dickinson, R. E., *The West European City*, London: Routledge and Kegan Paul, 2nd edn, 1962, 582 pp.

—, *City, Region and Regionalism. A Geographical Contribution to Human Ecology*, London: Routledge and Kegan Paul, 2nd edn, 1952, 327 pp.

Gottman, J., *Virginia at the Mid-century*, New York: Henry Holt, 1955, 584 pp.

Hobsbawm, E., *Nations and Nationalism since 1780*, Cambridge: Cambridge University Press, 1990, 249 pp.

12

Globalization and the New Territorial Order

After the Second World War, the transformation of economic life in train since the beginning of the industrial revolution, continued to result in a recourse to concentrated forms of energy (petroleum took an increasing place and the proportion of nuclear energy became important in the 1970s) and in a lowering of transport costs. But the information technology revolution brought a new dimension to change: linked to telecommunications it disturbed the economy of trade and of data processing. The speed of personal travel simultaneously facilitated contacts. The process of globalization, initiated in the sixteenth century, accelerated in the nineteenth century, was deepened and affected collective life in new ways.

12.1 Post-industrial Society and the New Territorial Order

The factors of transformation

In attempting to reconstruct the international economy after the Second World War, we must refer to the happy days of free trade established by the new institutions and to the economic aid plans necessary for the relaunching of economies ruined by the conflict. The productive activity of the 1950s put back into play a juxtaposition of national spaces similar to those of the 1930s. It was known, however, that they were no longer appropriate to the conditions of the modern economy: they were too narrow, especially in Europe, for it to be possible to gain all the external economies and economies of scale that specialization could engender.

Contemporary transformations of spatial organization scarcely result from the creation of common markets or large unified areas. The only example so far in this domain is the European Economic Community, which tends to distort our vision.

Progress in mobility is at the origin of the most spectacular mutations over the last thirty years. The use of energy continued to raise labour productivity and to restrict manpower needs in the primary sector of the economy. By the end of the 1950s, the secondary sector was in turn affected and workers decreased even if industrial production continued to increase. This gave rise to the idea of entry into a post-industrial era: manufacturing activities did not disappear but they ceased to employ the majority of the workforce. The tertiary sector was universally strengthened.

The use of the private car enlarged the areas drained by commuting which questioned the validity of the basic units of local organization. The units inherited from traditional civilizations, parishes or districts, were much too small to serve as frameworks for labour markets, even if they still had significance as neighbourhoods.

Rapid transport and telecommunications had equally spectacular effects, but at a different scale. Owed to them is the effacing of the nation as the privileged framework for productive activity. Without realizing it at the time, it was according to the costs of information transfer that the chains of manufacturing must be organized within the context of the national space, and it was due to the difficulty of diffusion of technical know-how that the old industrial countries had kept the monopoly of industrial work since the end of the nineteenth century. Within the space of a few years, generalization of education and new technology, founded on an easily transmissible scientific base, opened up a growing proportion of the world to manufacturing industry. Rapid transport and telecommunications permitted the management of distant plants or recourse to supplies of components or sub-ensembles located at the other end of the world.

States lost the power that they exercised over firms since they could always react to constraints imposed on them by relocating their production. Globalization of economic life made a rapid leap: no longer did it involve simply trade in raw materials and manufactured goods. It extended to the process of production itself. In each organization, the chain of manufacturing, assembly and finishing could now be widely dispersed over several states.

The new controllers of spatial organization: metropolization

The face of spatial organization was completely altered. The increased productivity in the primary sector continued to free labour and reduced the land area necessary to satisfy agricultural needs – progress that was accompanied by an unprecedented extension of pollution of rural origin. The cost of energy and raw material supplies was no longer a strategic location element except for certain heavy and primary processing industries; the reduction in maritime freight costs tended to locate them on littorals.

The constraints, which so far had determined the majority of locations for production, were reduced to very small significance: mines, fishing, agriculture and heavy industry, which remained linked to the distribution of resources, represented only a small percentage of total employment – less than 10 per cent in most advanced countries. The low transport costs moreover meant that a good number of plants gravitated towards the market. In the traditional world and since well before the organization of space resulting from the industrial revolution, the whole structure was controlled by the distribution of cultivated land that supplied the agricultural base of activity, and by the mines and energy sources. The relationship obviously was already loosened, which permitted an increasing part of the labour force and of activities to locate where accessibility to the market was greatest, in the central cores of nations, but the global structure continued to depend on the distribution of those whose primary activities forced them to live where resources were located.

In the aftermath of the Second World War, and with the exception of Great Britain who had sacrificed agriculture to free trade, farming retained over 10 per cent of the population – almost 25 per cent in France. In the space of a few years, the agricultural revolution induced by mechanization, the systematic use of fertilizers and progress in selection, swept all this away. Now only 2, 3 or 4 per cent of the workforce was needed to produce more food than could be consumed! Even by adding the services needed by this population, the primary distribution of population ceased to influence the whole structure of territories.

What counted from now on for industrialists and for services provided for firms – the significance of the latter had rapidly inflated – was to be in easy contact with a market extending to the entire world. The distribution of population was explained by activities at the top and not by reference to those forming the base. Telecommunications obviously gave access to world networks anywhere, but at unequal prices. When it was necessary to transfer huge volumes of information, location in a telecommunications hub benefiting from exceptionally low tariffs was a major advantage. For example this was the advantage of locating in New York.

Economic decisions imply partners who nowadays live almost anywhere in the world: when a new project is launched it is necessary to consult:

1. the firm's research and development section, often located far from the hubbub, in the outer suburb of a major agglomeration;
2. the commercial sections of client firms;
3. the managers of plants that the group owns (some may be located abroad) which will contribute part of the manufacture.

A telephone conversation is insufficient to resolve such difficult problems. Meetings are necessary, hours must be spent in negotiations, delays are made to collect new data, to consult other partners, then resume meetings. Only a location in a large city, well sited in relation to international airline networks is satisfactory. All space is organized from major metropolitan cities (figure 12.1).

The quality of life and spatial organization

The demands of long-distance communication are decisive in the articulation of contemporary life, but other factors come into play. Increased productivity is translated into less tense lifestyles for the majority of the population: the working day is shorter, weekends longer; holidays with pay are the norm. Working life starts later and ends sooner. People have become more sensitive to the quality of their environment, to the pollution affecting them and to opportunities for open-air activities, sport, shows and artistic events.

As the advantages of the environment one dreams of are generally not to be found near one's workplace, holidays are taken in tourist regions that do possess them. As free time increases so the importance of these areas grows in national economics. Today, it is better for a region to enjoy a pleasant climate, to be situated between sea and mountains, with towns charged with history and rich museums, than to possess mineral resources or fertile agricultural lands; Provence, which was for ages considered underprivileged, today attracts the masses.

People retiring from active life benefit from retirement pensions sufficiently comfortable to leave the place where they worked and to settle where the environment suits them better. This increases the weight of regions where life is pleasant. Often it is on the margins of the places where tourists flock during the high season and in environments that retain a rural charm, even if agriculture plays only a background role, that these new types of migrant settle.

Tourist zones are at the centre of intense movements, which leads to a multiplication of infrastructure investment: motorways and modern airports serve them. Accessibility is often so good that enterprises end up finding valuable conditions for locating there. Their workforce is content to be employed in pleasant surroundings. Medium and long-distance relations are easier than in non-tourist regions of equivalent

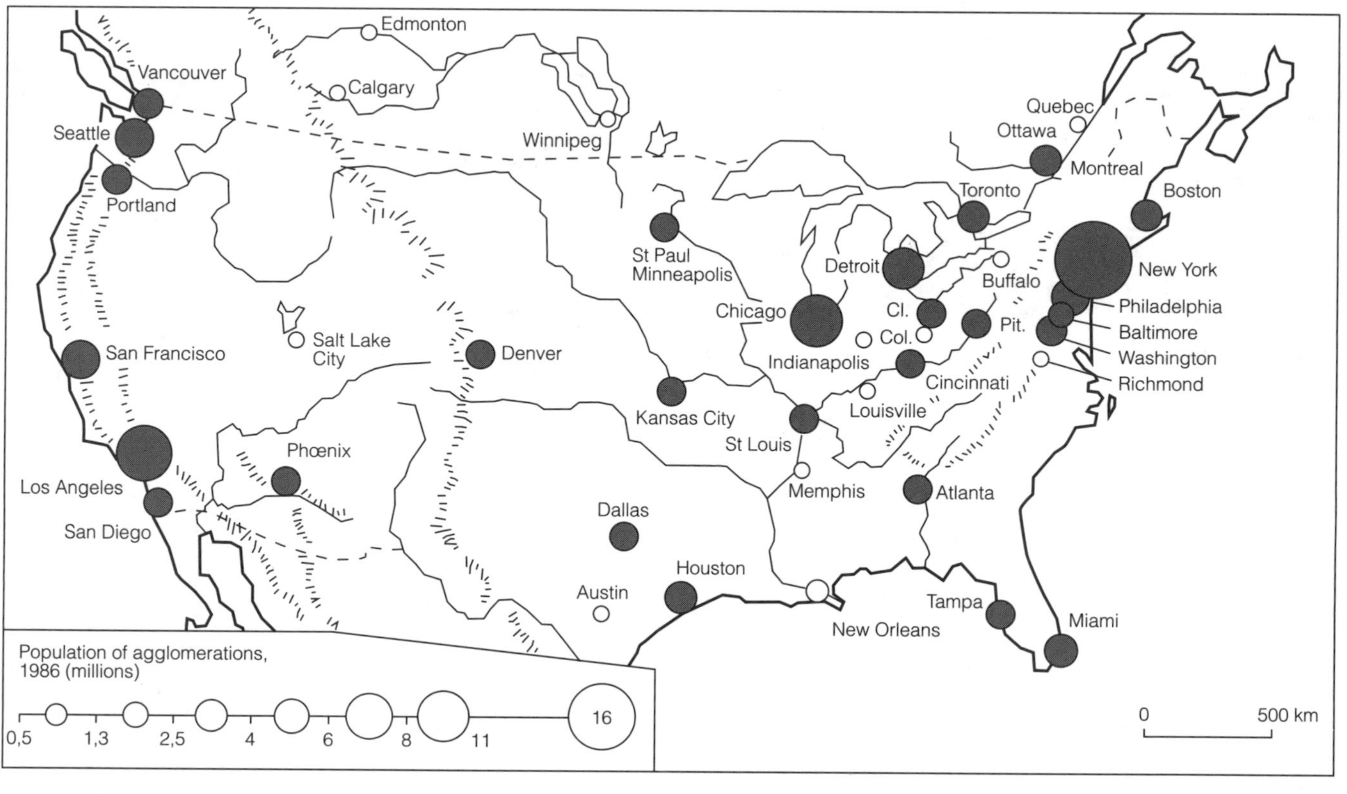

Figure 12.1 American metropolitan cities today

Agglomerations of over 1.8 million inhabitants, the threshold above which metropolitan cities may be defined today, are shown by shaded circles.
The density remains highest in the North East, but all the American regions possess a metropolitan city directly linked to the world economy.

Source: P. Claval, *La Conquête de l'espace americain*, Paris: Flammarion, 1989, p. 198.

population since their traffic justifies frequent services and prices are minimized by competition.

Regions where the pleasantness of the setting has for several generations attracted holidaymakers thus offer good conditions for the location of activities for people wishing to flee the congestion of badly-equipped urban centres that have been overtaken by modern technology. It transpires that these sunbelts, to use the expression that the Americans apply to the paradise of open-air life, possess metropolitan cities the sphere of influence of which is all the greater since they attract visitors from all over: one may cite Los Angeles, San Francisco, Phoenix, Denver or Miami in the United States; Barcelona in Spain; Rome in Italy; or Athens in Greece as being the largest. Medium-sized towns are also revealed as being attractive – Nice in France, or the Alpine towns.

12.2 The New Structure of Regions

The crisis of the old regional structures and the emergence of new forms of organization

The lowered transport costs and modern systems of communication removed the exclusive advantages enjoyed by the central areas of nations. From now on these favourable conditions were shared by all the metropolitan zones directly connected to the world network of rapid contacts. They attracted the services demanding contacts with distant partners and manufacturing activities that had these same needs. Others located in centres well served by a second-level airport, close to a regional capital for example. An increasing number of activities located anywhere they wished, swelling the current of counter-urbanization, which over the last twenty years balances the forces tending towards metropolization.

Specialization of firms was still imposed by competition, but depended less than in the past upon agriculture as industry depended less on natural endowments; the firm size and the human and capital resources at its disposal counted for more, to the extent that in the same areas of economic activity the tendency is for a mixture or juxtaposition of different sectors. The inherited territorial structures, in countries of ancient inhabitation, in traditional societies undisturbed by the industrial revolution except for the simple accentuating of the hierarchy of centres and the role of regional capitals, were less and less valued. The lowering of densities in a major proportion of rural space and the total disaffection found where the resources are poorest, reduced the lower ranks of the urban networks to practically nothing. The smaller regional capitals, or those that it was not easy to equip with rapid means of movement towards centres at an international level are losing ground. Contrasting with emptying zones is a marked demographic expansion in all sectors where the quality of life is pleasant. Threshold effects appear: above 200 inhabitants per square kilometre it is possible to provide services of an urban character to a dispersed population. With lower totals, a linear regrouping along a few corridors leads to the same result. Below twenty inhabitants per square kilometre, services become so mediocre that the majority of people prefer to live somewhere else.

The basic units of collective life are larger than in the past: due to the private car and public transport, labour catchments cover impressive areas, 100 square kilometres at a minimum and sometimes 1,000, 5,000 or more. The daily frameworks of existence are extensive: the 'localities', in the sense used by the British geographers, are vast and their structure complex, ordered by one or several poles of activity, interleaved by virtue of the diversity of residential zones and controlled by the hierarchy of commercial or recreational centres. Regional organization combines highly contrasted elements. One may distinguish:

1 zones of abandonment, which are returning to nature or serve certain types of recreation;
2 areas or primary resource exploitation, with densities so low that young adults complain of the remoteness of the services they would like to enjoy;
3 rurban or suburban areas, where agriculture provides a modest proportion of incomes and where retired persons, people employed in teleworking, and workers or managers employed in dispersed firms located in the nearest urban zones are found;
4 urban zones, articulated at three levels, capitals at an international level, cities which are well connected to them, and satellites of one or the other.

The configurations of the new forms of territorial organization lack the geometrical regularity that were noted with satisfaction in bygone days. The restructuring implied by their development is very substantial. It is marked first of all by the ruins of the old industrial region, victims of competition by the new producers. In the United States, the classic contrast between the Industrial Belt and the entire peripheral zones is in the process of being removed. The crisis of the old activities reduces the advantages of the old manufacturing core (we speak now of the Rust Belt). It only maintains its position thanks to the high technology industries which are now found in all the country's metropolises (figure 12.1): they constitute the driving force in the new system. The distribution of cities of two million inhabitants with direct access to long-distance connections remains more dense in the old north-east, but all parts of the Union possess them now: Miami, Atlanta, Houston or Dallas-Fort Worth in the south, Denver, Phoenix, San Diego, Los Angeles, San Francisco and Seattle in the west. The growth of these centres is more rapid in the Sunbelt, where they benefit from a more pleasant climate and a better quality of life.

The old centre–periphery model has lived out its life. The functions of innovation, stimulation, control and management have ceased to be the privilege of the zones most accessible by surface transport within a relatively closed economic space – whether of nations, continental areas or the world. These functions are now devolved to the major metropolises linked easily by air transport whatever their distance might be. The massive central cores of the past have been replaced by multiple poles, often widely separated from each other. They directly control the activities of zones served by second-ranking airports several hundred kilometres in radius – important provincial towns, past regional capitals often, as well as suburban and other rurban zones, or small towns within one hour's radius of their airports. The rest of the space is divided between low density zones of primary resource exploitation and areas of human and economic abandonment and reversion back to nature. The 'peripheral' spaces are just as splintered as the central areas. The possibilities offered by telecommunications and teleworking favour counter-urbanization, but these only profit areas with good quality services. They remain limited to rurban zones with relatively high densities that are well served by rapid communications.

Spatial organization is more marked by the progress of communications than in the past: structuring by networks is more important and limits the importance of factors of homogenization, whether they are ecological, ecologico-economic (as was the case in the specialized zones in the first phases of the industrial revolution), social or cultural.

The contemporary mutation of the forms of territorial organization often lead to a better distribution of the cores which impel and direct economic life, as in the United States, but such an evolution is only possible where the population density and totals are sufficient: in Canada for example, the only nodes to be properly connected to the world network remain Toronto and Montreal, both above the threshold of two million inhabitants characteristic of the North American metropolises. Vancouver approaches this and thus possesses long-distance connections that give it an opportunity in the new competition. By contrast, this is not the case in the maritime or prairie provinces, where activity remains essentially limited to the exploitation of natural resources, which requires ever fewer hands.

Within the European Economic Community, the change in locations has avoided, as many feared, the concentration of population and activities on the central axis from London to northern Italy. De-industrialization has been just as severely felt in the 'black countries' within the axis as in the more peripheral areas. Political fragmentation ensured that all the capitals had an occupational differentiation and rapid communications which classified them at the level of metropolises. In certain countries, major industrial areas, or service centres (in the latter case, often less populous, like Geneva) enjoyed similar advantages. All the countries have thus benefited from growth, but the gains were most apparent in the European sunbelt than in northern Europe; all have equally experienced the global development of their cities, surrounded by aureolas of medium density or small towns. In all the depopulated areas one finds an increasing dependence on primary production and on the existence of abandoned areas. The relationship of these diverse types of spatial organization varies from one country to another, but none is excluded from the general dynamic.

France possesses, in Paris, one of Europe's major capitals (figure 12.2). Two other zones also enjoy good accessibility to the world as a whole: Nice and the Côte d'Azur, and the French portion of the agglomeration of Geneva (a zone which extends from Morez or Saint-Claude, in the Jura, as far as Annecy, passing by the Gex region and the southern shore of Lake Geneva). The growth of these areas has been constant despite recession. Ever since airlinks became dense, cities over 400 kilometres distant from Paris are more accessible in time–distance than those that are located closer: it took more time, before the *Train à Grande Vitesse** was opened, to travel from Paris to Lille than from Paris to Nice, Marseille or Toulouse. It is thus the major cities of the French periphery, from Nantes to the Côte d'Azur, to Lyon and the Alps that have experienced the most rapid development in the course of the last fifteen years. Medium-sized towns like Pau, Perpignan or Montpellier have managed to extricate themselves, Strasbourg and Alsace equally cannot complain. The Paris basin, which benefited from the majority of industrial decentralizations linked to the policy of controlling new implantations in Paris between 1955 and 1970, did not experience the same dynamism, even though its central part benefited from counter-urbanization.

The territorial organization of advanced countries is far from being frozen. The aeroplane has given opportunities to all the metropolises and has caused the advantages of central zones to disappear. But high-speed trains could permit them to recover part of the lost ground: this is what is seen between London, Hamburg, Florence and Paris

* The French high-speed train system, which can attain a sustained speed of 300 kph.

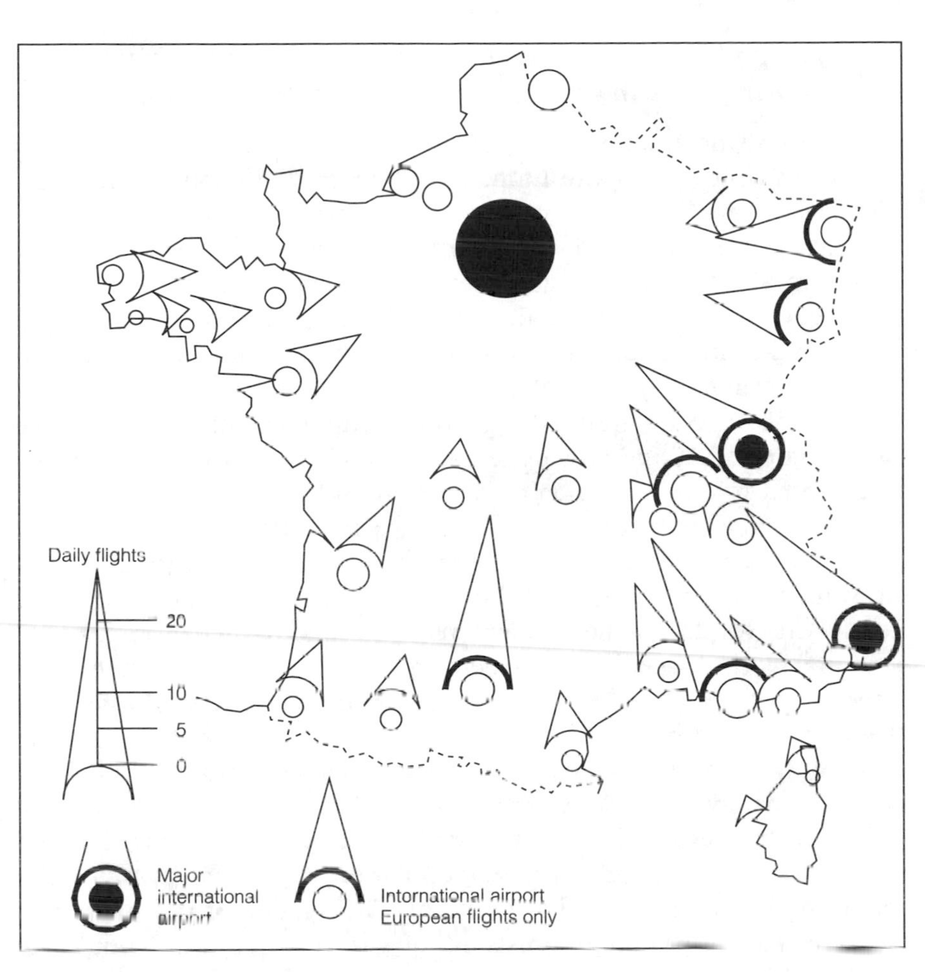

Figure 12.2 Metropolitan cities and regional change in France
The evolution of major cities is today determined by their contact facilities, for example, their airline services, which provide international connections for world ranking metropolitan cities (the enviable position of Geneva and Nice can be seen from this point of view), connections with Paris for the second-level metropolitan cities (towns within 400 km of the capital are disadvantaged, even including the impact of the TGV).
The circles are proportional to the size of towns, the arrows indicate the number of daily connections with Paris via Air Inter (September 1992).
In black, the agglomerations benefiting from dense international connections.

where the completion of systems, and those that are programmed, will bring closer together the centres that aircraft serve increasingly badly because of the progressive congestion of airspace. Lille, one of the nodes in the European high-speed train network finds itself in an enviable position.

The problems of advanced countries: the restructuring of metropolises, pollution, the loss of power of the state

Within these new spatial configurations, the metropolises are in a state of complete reorganization. The former radio-concentric plans have given way, due to the private car, to structures both freer (they have an enlarged radius) and more massive (they have ceased to grow as fingers along railways). Their service depends on motorway networks and public transport: the former lend themselves to multiple origins and destinations such as suburb to suburb; the latter are essential for movement towards the centres if they are not to be allowed to succumb to congestion. In the most important metropolises like Paris (figure 12.3), growth benefits the original centre, which keeps an essential role in the fields of direction and recreation, head office complexes like those of La Défense or Bercy, the suburban sectors which attract office or research activities (around La Défense, or in the southern technopôle), and the major airports (Orly and especially Roissy-Charles-de-Gaulle), which ensure access to the national, European and world space. The day seems not far away when the two complementary circulation systems will be saturated. The multiplication of underground connections will improve the situation but, in the long term, only a more systematic recourse to telecommunications can avoid the paralysis of these gigantic organisms.

The central regions have lost a proportion of the advantages which led to the accumulation there of population and activities since the beginnings of the industrial revolution. The decline of old industries has reduced the polluting emissions of gas and liquid effluent. The pressure exerted by post-industrial societies on their environment has not, however, diminished: the raising of living standards and mobility has increased the consumption of energy and pollutant waste. The resulting problems, which also arise from agricultural intensification, now have a regional, even international, dimension: areas affected by direct atmospheric pollution, whether urban or industrial, cover thousands, sometimes tens of thousands of square kilometres; indirectly, their effect is felt in the form of acid rain on a continental scale. Ecological disequilibria give, both via the harm that they engender and the costs necessary for the treatments that they necessitate, economic constraints which in many cases makes the pursuit of the concentration of activities difficult. The dynamism of metropolization is, however, so strong that in a country like Japan, it was the absolute growth of the Tokyo agglomeration, already reaching 30 million inhabitants, that was the strongest in the 1980s.

The weakening of states results firstly from the new facts of economic life: enterprises are no longer constrained by difficulties of

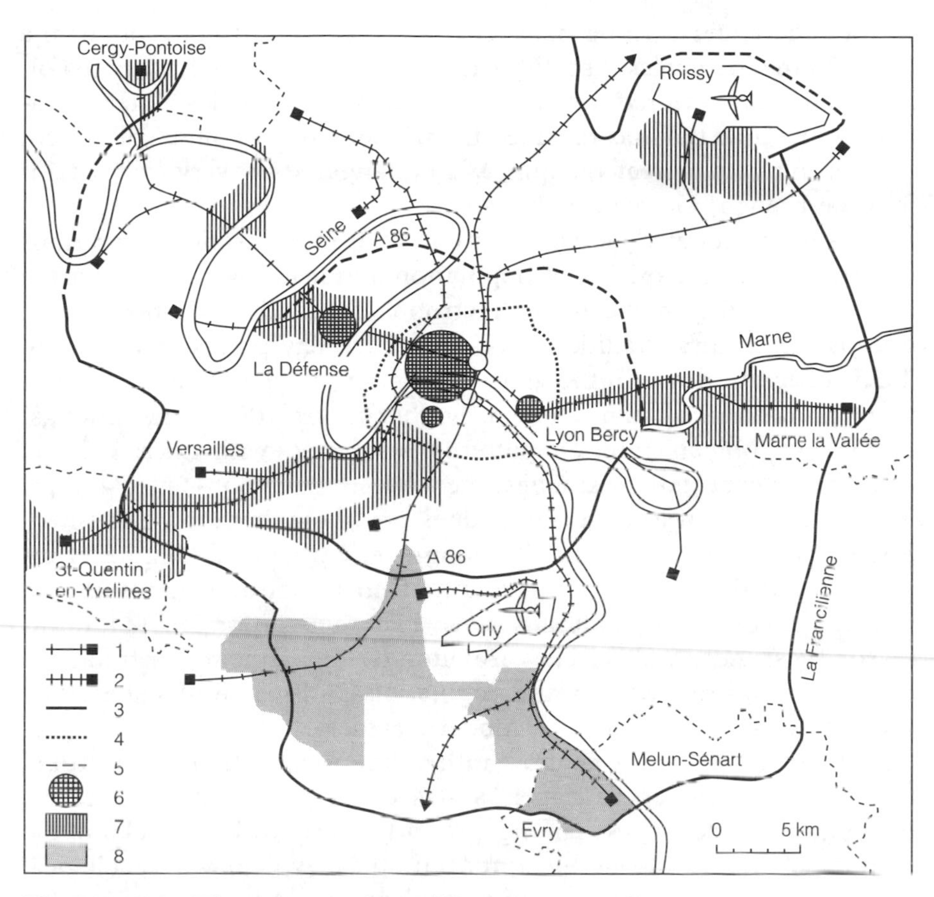

Figure 12.3 The restructuring of Paris

1 Lines A and B of the RER and terminal stations
2 Lines C and D of the RER
3 Ring motorways
4 Paris city limit and the peripheral boulevard
5 Boundaries of New Towns
6 Headquarters and business centres
7 Sectors of rapid development of dispersed office activity
8 Technopôle

Paris has become a polycentric city in which the zones of new activity are linked to airports and rapid routes (RER and ring motorways) which structure the agglomeration.

communications to concentrate the essential of their production chains in the same national space. They can thus respond to the measures that governments try to impose on them by locating part or all of their production elsewhere. In parallel a decline is observable in the ideologies that cemented national consciousness. In a world dominated by mass culture, the trend is towards uniformatization and a weakening of the differences existing between states. The elitist concepts of the past and the ideologies that they carried, gave a sense of history to collective life, and ensured the cohesion of national societies. The specialized know-how that has replaced them, today is very often splintered, thus opening up a great vacuum. We no longer know very well what it is to be French, German or Italian.

This is a particularly serious matter when long-distance migrations diversify the ethnic origins of population and lead, in urbanized areas and in particular in the major metropolises, to the juxtaposition of groups the cultural traditions of which have few points in common. Their integration within the global community appears to be increasingly problematical. Simultaneously the general diffusion, amongst the young, of consumption patterns associated with mass culture, creates high expectations amongst the adolescents in non-native groups. Their qualification level generally does not allow them to be satisfied. Such a situation favours the development of a general climate of insecurity and the outburst of unruly movements of destructive anger.

The ideologies of modernity have lost their attraction – for this reason, post-industrial societies are often termed as being postmodern. An important proportion of the population is in search of values. The great religions reply badly to modern anxieties: have they not professed for more than a century rather uncritical attitudes regarding technological progress? Chapels and sects respond better to needs. Amongst the ideologies gaining ground, those which proclaim the divinity of nature are at the forefront: they explain why ecological movements gain favour and the new hierarchy of places that results. In a world where the disappointed utopians can be counted in millions, the nostalgia for traditional society and the pre-industrial world counts for much: it is at the heart of the renewal of regionalism, in certain cases of nationalism, and of the concern to preserve as much as possible of the landscapes linked to the activities of the past.

Decline and resurrection of territorial sentiments

Postmodern society reinvents, in a way, means of creating a hierarchy of the real world and of conferring different ontological laws on places. Attempts in this domain are numerous: it is difficult to distinguish clear dominant ones.

In the traditional world, the general low level of mobility and the vigour of local cultures gave much meaning to being rooted: the *pays* where one had spent one's childhood was where one felt at ease and where one escaped the prejudices accorded to strangers. If one felt French, it was firstly because one was Breton, Savoyard or from Lorraine.

The modernization of societies has radically modified these conditions over the last two generations: in a country like France, sentiments had not greatly changed at the eve of the Second World War outside the industrial regions and Paris, where the mixing of populations had long since weakened traditional forms of organic solidarity. Today, it is at the scale of the evolved world that they disappear. Mobility has increased: the part of the population that is simply transitory is high and the movements of individuals take place within a widening framework. Popular cultures have lost their strength. The generalization of compulsory education had already affected them by the end of the last century. They disappeared under the effects of the extension of schooling and the increased role of the media.

This banalization of space is displayed in many signs: France has lost its Midi and now only has a South. Local and regional attachments become weak as a consequence. But in the world without values that we have entered, putting down roots (chosen and proclaimed rather than endured) has become a fashion and permits a group to have an identity and lay claim to its traditions. The traditional region has been emptied of a part of its authentic content, but it is resurging like a phantom and as an object of cultural consumption. This is one of the significant facts of the political geography of the present-day world.

The sequence that we have evoked of forms of spatial organization which have succeeded each other since the beginning of the industrial revolution characterizes the first countries to have entered the race for progress and to have accepted the simple rules of market trading. Elsewhere the evolution has followed different paths. For a long time, the trends seemed to diverge. A certain convergence is observable today.

12.3 The Third World: Other Sequences of Regional Evolution

Globalization of the economy and territorial organization

On the eve of the Second World War, world opinion became aware of the problems posed by the widening gap between the economic level of the industrialized world and that of other countries. It was the epoch when techniques of national accounting became widespread: for the first time, data permitting a comparison of average income per capita of Americans as compared with Peruvians, Afghans or Haitians became available. As account is not taken of the non-monetary household economy, the figures exaggerate the differences. In spite of these imperfections, or because of them, they have a striking effect: how can one live with a clear conscience when one is part of a democratic society with egalitarian ideals when one knows that there are peoples with per capita incomes ten or twenty times lower than one's own?

The world discovered the problems of unequal development. One began to talk about the 'Third World'. This grouped together countries with very diverse cultures, modes of social organization, densities and resources, but all of them in the 1950s suffered from high birth rates which caused their populations to explode: it was an effect of the revolution of mass medical provision, which, in a few years, had permitted a halving of mortality rates. In the face of this demographic dynamism, investments were too low to assure rapid take-off. Economists, who understood how national economies functioned and the advantages of regions enjoying the best accessibility to the entire market, discovered the logic of centre–periphery relationships and declared themselves very pessimistic about the future of countries now referred to as being in the process of development.

These theories were already dated in relation to the reality. The increasingly scientific character of technology and the revolution in rapid transport, telecommunications and information technology, from 1955 or 1960, broke the locks which manacled development. It was progressively easier to find in the Third World, labour supplies and the technical and administrative skills demanded by manufacturing industry. Cables or satellites provide instantaneous links between the plants that multinationals have chosen to locate in countries where labour costs are derisory, and with head offices in London, Paris or in a major agglomeration in the United States. The areas placed within an hour of a major international airport were the first to benefit from implantations, but others followed in satellite towns a little further away or in large regional towns with good airlinks.

In the space of a generation, the liberal economies of south-east Asia, from South Korea to Thailand, to Malaysia, to Indonesia, became the land of new industrialized countries. Its major metropolises (Seoul, Taipei, Bangkok, Kuala Lumpur or Jakarta) are well linked to the entire world. Singapore and Hong Kong are two of the several driving and control centres of the world economy.

The Third World is diverse, but the countries that compose it have in common forms of spatial organization that do not resemble those born of the industrial revolution in Europe. Rural population often remains high, in spite of the strength of urbanization, but remains too poor to attract the location of new activities: the latter are located from above, that is with the possibility of long-distance relations. The Third World is living in the age of metropolization. Around its most dynamic agglomerations, secondary centres are in turn caught up by the dynamism, but the diffusion ends quickly. At 50, 80 or 100 kilometres from the cities, the bitumen cover of most roads comes to a halt. Instead of driving at 80 or 100 kilometres per hour, with the slightest rain one slaloms between the ruts and is happy to cover twenty kilometres in an hour! The universe of high mobility in modernized areas is left behind.

The example of Brazil

Brazil offers an excellent example of the forms of spatial organization in what thirty years ago was termed the Third World. The term is no longer applicable to a country that belongs to the newly industrialized world. The modern economic sector occupies more than a third of the population and industrial production exploits the most advanced technologies. The traditional life is organized by its relationship to the ports where exports of raw materials take place and manufactured articles arrive – a slave and freeman relationship. The interior supplies a not insignificant portion of the food supplies for the exporting zones of the littoral by selling it its livestock. An entire territorial organization, tenuous but effective, was established from these penetration routes, equipped with trails initially and railways later.

Modernization was preceded at the end of the 1940s and during the course of the 1950s by the construction of a road network which unified the national market. From then on, growth was organized around the metropolises each of which dominated a large part of the country, or assured its global organization and its link to the outside world. The upper level (figure 12.4) is articulated around Brasilia which effectively took over the political direction of the country from 1965; from Rio de Janeiro, which has the best-served airport, dominates the information and communications networks (with powerful national television channels in particular) and houses the headquarters of nationalized firms; São Paulo which controls all economic and financial life. The second level is that of major coordinating centres. The most active count over two million inhabitants: Porto Allegre, Belo Horizonte, Salvador and Recife. Another generation of metropolitan centres is already in place, with populations close to or exceeding a million inhabitants: Curitiba, Fortateza, Belem or Manaos are good representative examples.

Almost all the Brazilian space today (Amazonia still being exceptional) is provided with a modern transport and communications network which makes Brazil a national market and articulates it with the international world. The essential problem, obviously, for such a country is that of the cultural, economic and social advancement of the population which remains in misery.

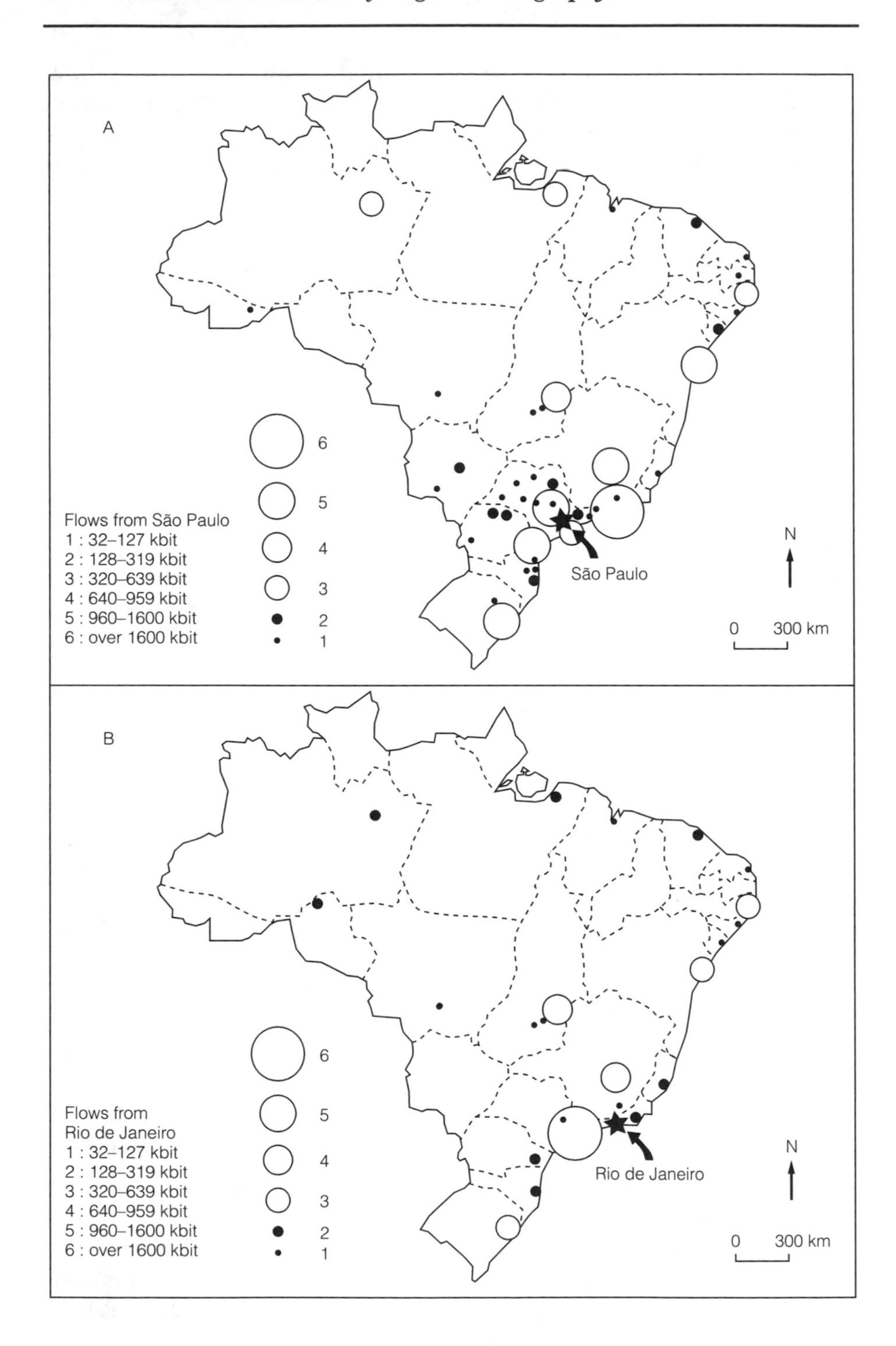
A
6
5
4
3
2
1
Flows from São Paulo
1 : 32–127 kbit
2 : 128–319 kbit
3 : 320–639 kbit
4 : 640–959 kbit
5 : 960–1600 kbit
6 : over 1600 kbit
São Paulo
N
0 300 km
B
6
5
4
3
2
1
Flows from
Rio de Janeiro
1 : 32–127 kbit
2 : 128–319 kbit
3 : 320–639 kbit
4 : 640–959 kbit
5 : 960–1600 kbit
6 : over 1600 kbit
Rio de Janeiro
N
0 300 km

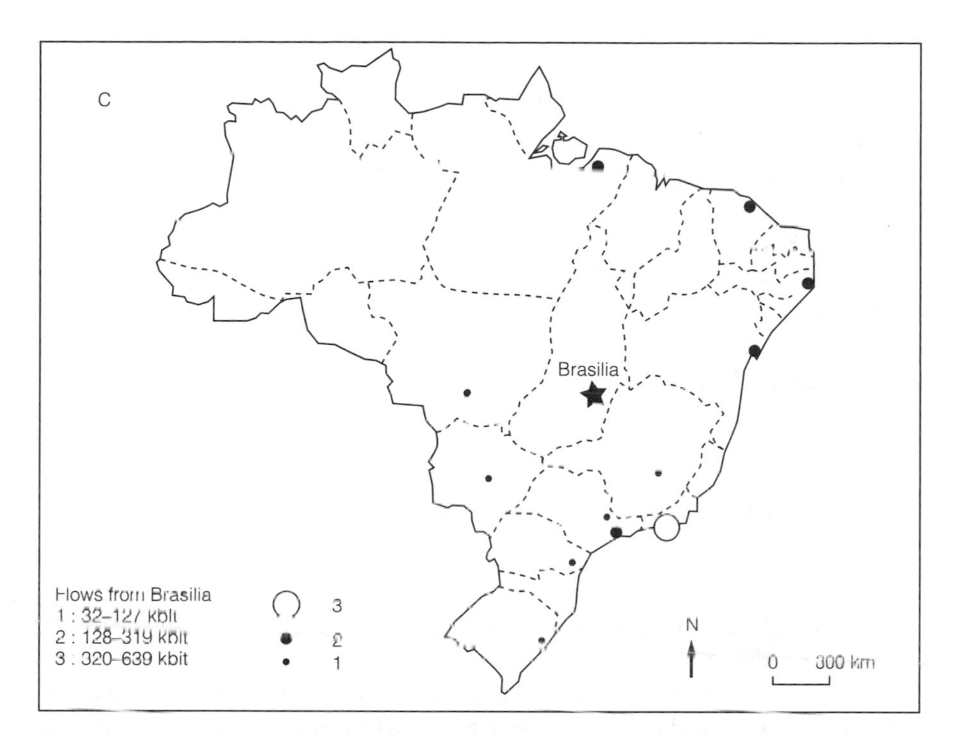

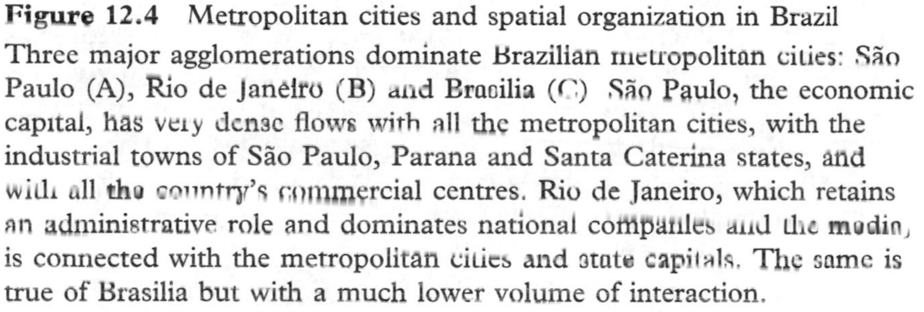

Figure 12.4 Metropolitan cities and spatial organization in Brazil
Three major agglomerations dominate Brazilian metropolitan cities: São Paulo (A), Rio de Janeiro (B) and Brasilia (C). São Paulo, the economic capital, has very dense flows with all the metropolitan cities, with the industrial towns of São Paulo, Parana and Santa Caterina states, and with all the country's commercial centres. Rio de Janeiro, which retains an administrative role and dominates national companies and the media, is connected with the metropolitan cities and state capitals. The same is true of Brasilia but with a much lower volume of interaction.

It is the new industrial countries of non-communist south-east Asia which have travelled furthest down the road of the integration of the totality of their population in a territorial, social and productive modern organization: the mutation has been achieved in South Korea and Taiwan and is well advanced in Malaysia. It is progressing rapidly in Thailand and in Indonesia. In the latter two countries, delays and gaps are significant, but less than in the Philippines. In the case of city states like Hong Kong or Singapore, modernization has not made the same efforts in the social domain. It is there that evolution has taken the most spectacular forms.

Not all the countries of the Third World have achieved their take-off, but even in those the economies of which remain agricultural or based on minerals and where incomes are stagnating, the tendency towards metropolization and the development or regional structuring close to that developed thirty years ago in the industrialized world is asserting itself.

The causes of failure in the race for modernization and of the restructuring of space that is called for are multiple. The mediocrity of resources and recurrent years of catastrophic drought have blocked development for the last twenty years in the Sahel countries and the Horn of Africa. Political instability and armed conflicts have ruined, or are ruining, countries like Nicaragua or El Salvador and, to a lesser degree, Colombia and Peru in Latin America; Sudan, Ethiopia, Somalia and Eritrea in Africa; Lebanon and Afghanistan in Asia. The situation remained the same for a long time in Mozambique, Angola and in Vietnam.

12.4 Globalization and the Failure of the Socialist Model

The social fragmentation of countries where tribal structures remain active, and the exacerbated dualism of some where peasant agriculture persists, also put a brake on growth. The essential evolutionary factor, however, rests in the political orientation chosen at the time of independence or later. To break with the hated West, the fashion was to copy the Soviet model, or in the case of the Muslim world, versions tinted by the Koran. The results are variable, but they are generally bad, even catastrophic: the agricultural bases have been ruined without industrial capacities being sufficiently developed. The infrastructures have stagnated and urbanization is not being accompanied by significant sociological changes.

The socialist world has for a long time perpetuated illusions. The Second World War imposed socialism on Eastern Europe. The Revolution inserted China into the communist camp in 1949. The recipes proposed by the USSR were applied, without soul-searching, to countries with very diverse traditions and levels of development. Was it not necessary everywhere to increase the production of foodstuffs, energy and minerals? Was it not essential to accelerate industrialization – or to encourage it where it was absent – to catch up on the delay behind the capitalist world and to overtake it one day?

In 1945, the birth of a socialist international space could have been contemplated: the USSR was no longer the only nation to have changed its system; an opening up seemed possible. This did not happen. For this to have been achievable, it would have been necessary for the Soviet Union to renounce the system of centralized planning and return to price-fixing by the market, which was ideologically impossible. After 1955 when the East European countries launched the experiment of

COMECON, which was supposed to be the counterpart of the Common Market, they lacked flexible national adjustment policies to permit their economies to be truly integrated.

For forty years, the chosen orientation remained essentially the same in spite of the increasingly obvious signs of exhaustion; the socialist system had in effect an inbuilt fundamental contradiction. The progress that it achieved at the outset by a more effective exploitation of resources created an increasing demand for communication which it had not the means to satisfy. The reforms launched by Khrushchev did not attack the fundamental vice of the regime. This remained the pursuit of a national model economy which long ago had ceased to characterize the countries it was determined to imitate and overtake.

It was definitively the refusal of globalization that condemned the socialist economies to failure: they were shown to be incapable, in spite of the spectacular success of their military production and their space policy, to increase the living standards of the population, to put in place a satisfactory infrastructure of communications and services, and to supply the health needs of a population made more demanding by its progress in education.

In these countries, the new forms of production have well and truly flattened the features of traditional spatial organization: the compartmentalization of the former rural field system had been overtaken by excessively large land blocks resulting from collectivization; for the dispersed farms, hamlets and villages, it was attempted (with varying success) to substitute agro-towns; the urban centres lost a significant part of their tertiary functions at the time when these were increasing in the West. In a sense, socialism led to a real regional destructuring.

As far as mentalities were concerned, the mixing of populations was weaker than in the West – except in the industrial regions. The traditional inward-looking and chauvinist attitudes were maintained. Local and regional feelings, not to mention ethnic and nationalist movements, found in this their best nourishment.

The collapse of the Berlin Wall suddenly revealed the gaps in a system that was insufficiently interested in places to protect them, and in space, to provide it with the means to master distance and allow all its members to participate in the same general communication networks.

The collapse of the East European communist countries confirmed the inherent weaknesses in their economic and political systems. By turning their backs on the market, they were condemned to live an inward-looking existence at the time when the dynamism of market economies was stimulated by the globalization of production chains. As they tried to develop more sophisticated industries, so the exchange structure of their information was shown to be more defective. By the beginning of the 1980s, it had become evident that these regimes were incapable of catching up and overtaking the liberal economies. The present crisis of Soviet communism simply confirms what all good geographers have known for a long time: the centralization of decision-making circuits generates so many dysfunctions that modernization quickly hits impenetrable barriers.

Conclusion

Since the beginning of the industrial revolution, the forms of spatial organization have shown for a long time a divergence between the developed centre and the periphery on our planet. In the last thirty years, we have entered a phase of convergence linked to the revolution of rapid transport and telecommunications: all the areas close to metropolises now belong to the central core of the world economy; all are experiencing the same problems of immigration and cultural integration. The peripheral zones have also fragmented. Even if their settings resemble each other, the problems are not always expressed in the same way. In developed countries, it is the attacks on the environment and the problems of cultural cohabitation that receive most attention. In the East, it is the transition from structures deliberately imitated from nineteenth-century Europe to the organizational forms of the twenty-first century that must be achieved in the next ten or fifteen years: an efficient agriculture, which respects the environment, must be created; metropolises must be given the means which they lack to be fully integrated in the world network of rapid communications, resolve the worst problems of pollution and plan the reconversion of the old industrial zones. In the majority of the Third World countries, the priorities concern employment and basic equipping, but pollution is often severe and demands urgent action.

The convergence of modes of territorial organization is evident simply by considering the zones of urban concentration and the manner in which relationships are organized around them. But a new element is the abrupt rise of anti-western ideologies: fundamentalisms have ceased to be marginal phenomena. They question the perception of space, the methods of exploiting it and the priorities that are imposed upon it.

The forms of regional organization of bygone days are often invested with new values. Should they be simply revived? Nothing is less certain. The world is organized more and more as a function of multiple networks which facilitate trade, communications and mobility, and liberate people from the limitations that obstructed them when they lived in narrower frameworks: we have seen, since the time of ancient societies, contact structures asserting themselves. We have passed from networks that were so tenuous and so unstructured that their mark on the landscape was barely visible in an increasingly hierarchized mesh. The evolution has long been towards a concentration of control in capitals and for that of economic activities of stimulus and direction into the central areas of national spaces. The present-day evolution of networks is marked simultaneously by their strengthening and restructuring which for the first time and at all scales has made polycentric organizations with multiple heads effective.

The inhabited space is increasingly one of extended networks: instead of being contained in the narrow circle of movements on foot, the interweaving of daily movements and contacts are woven at the scale of localities of several hundred or several thousand square kilometres. The circuits of economic decision making are set at a world scale. The logic of administration and territorial control maintains, in political life, a hierarchy of authority, but their highest levels are tending to strengthen, which conforms to the general direction in the framework of relationships: what arises from international cooperation or supranational authority is increasingly important. This comes back to the point that many decisions are taken within networks that structure much larger spaces than in the past.

At the same time, and it is a specificity of policy, the local level is experiencing a renewal of interest. This results from the normal evolution of societies and economies which have become so complex that it is necessary to avoid the congestion of their higher levels of authority. But the renewal of local feeling and the nostalgia for relatively closed forms of spatial organization have other causes. The identities intimately linked to the feelings people hold of their lives can lean on the solidarities that go with local networks of social links, as well as with those born of more distended connections. The former for their part have a feeling of being more immediate and to seem more natural, but they cohabit with others once the basic units have been opened up. Large identities have the specificity of finding justification in universal religions, transcendental metaphysics or the progressive philosophies of history.

These forms of thought are in crisis today. Western thought no longer believes in its religious and philosophical foundations and we live, as Francis Fukuyama has said, in the end of history. It is this vacuum that opens the way for a renewal of tribalism and fundamentalism, and for the rage of certain nationalisms.

The regional question is not, in the present world, a minor problem: it expresses the disarray of people in the face of the organizations of space that they have taken centuries to build and the meaning of which seems today to escape them.

Further Reading

Brotchie, J., Batty, M. and Hall, P. (eds), *Cities of the Twenty-first Century: New Technologies and Spatial Systems*, London: Longman, 1991, 446 pp.
Castells, M. and Hall, P., *Technopoles of the World*, London: Routledge, 1994, 320 pp.
Cole, J. and Cole, F., *The Geography of the European Community*, London: Routledge, 1993, 317 pp.
Dicken, P., *Global Shift: The Internationalization of Economic Activity*, London: Paul Chapman, 2nd edn, 1992, 492 pp.
Franklyn, H., *The European Peasantry: The Final Stage*, London: Methuen, 1969, 360 pp.
George, P., *France: A Geographical Study* Part Three, 'New Regional Structures', London: Martin Robertson, pp. 133–222.
Gottman, J., *Megalopolis: The Urbanized North-eastern Seaboard of the United States*, Cambridge, MA: MIT Press, 1990, 249 pp.
Hall, P. and Preston, P., *The Carrier Wave: New Information Technology and the Geography of Innovation, 1846–2003*, London: Unwin Hyman, 1988, 305 pp.
Hepworth, M., *Geography of the Information Economy*, London: Bellhaven, 1989, 258 pp.
Kellerman, A., *Telecommunications and Geography*, London: Bellhaven, 1993, 230 pp.
King, A. D. (ed.), *Cultural Globalization and the World System: Contemporary Conditions for the Representation of Identity*, London: Macmillan, 1991, 184 pp.
Planhol, X. de and Claval, P., The France of large organizations, ch. 11 in *An Historical Geography of France*, Cambridge: Cambridge University Press, 1994, pp. 415–68.
Scott, A., *New Industrial Spaces*, London: Pion, 1988, 132 pp.
Shaw, G. and Williams, A., *Critical Issues in Tourism*, Oxford: Blackwell, 1993, 304 pp.
Smith, D., *The Ethnic Revival*, Cambridge: Cambridge University Press, 1981, 240 pp.

13
Conclusion

13.1 The Region and Progress

Geography was born of the regional approach, from the review that the traveller makes of his impressions, to dominate them, to organize them and to interpret them by placing them on a map. But it is not sufficient to review what one has seen to discover worthwhile realities. Observation must be educated and the mastery of analytical disciplines is indispensable. They allow the understanding of the influence of ecological, economic, social and cultural forces which combine in the structuring and organization of space. They reveal the influence of politics that the organizing of space always implies. They show, finally, how the intimate being of social groups shapes itself to the constraints of the environment and identifies with the *pays*, region or nation where they live.

Beyond this analysis, regional geography invites a reflection on society. Human groups are not constituted arbitrarily, in a Platonic Universe of Ideas imperfectly reflected in the world. They evolve in a material environment which is never indifferent to the awareness they have of themselves, offers them the setting where they live and supplies them with food, energy and the materials they need to live and carry out their activities.

The imperatives of food production for a long time constrained people to dispersion and obliged them to practise ways of life adapted to the environment where they were settled. Have these constraints hindered the development of social contact? Yes, from several points of view: the life of contacts is necessarily limited when dispersal is so total that one never mixes with crowds and one only meets partners that one knows too well and for too long. But the disciplines imposed by the mastery and exploitation of the environment have taught individuals to assume coherent rules and to structure themselves in social groups. Through an extremely limited spatial experience, they have made the transition from a fragmented humanity to universal categories. Regional constructions thus appear as being much more than a material background: they inform us as to what has conferred stability on groups, on the way in which they were able to insert themselves in the natural world and tie permanent links with it, and of the idea they have of themselves.

The adventure of modernization is extraordinary. Little by little, people have learned to overcome local constraints. They have become more mobile and have achieved the possibility of the exchange of information even more easily and over greater distances. The goods they produce from the ecological support that the earth constitutes have not ceased to diversify. Groups have freed themselves from dependence on the local resources where they lived. They formed relations that extended from the enlarged locality to the entire planet. For the dispersion to which they were previously condemned has been substituted a large choice of locations, with the possibility of living in megalopolises where the variety of interactions offered is prodigious.

After the time when space was organized from below, by mastering local ecological pyramids, came the epoch when everything is ordered from above, according to accessibility to the world space. We tend to forget what all this implies, for the gains are not without losses. These are to be seen, above all, in the increasingly massive dysfunctions that human groups introduce into ecological mechanisms. They are also apparent in the inability of enlarged societies, concentrated in pockets of density of previously unimaginable magnitude, to provide their members with jobs, a roof over their heads and a reason for living and hoping. The defects and disequilibria of cultures today are apparent in the unfinished and brutal character of the environments they fashion.

To interrogate the way in which people inhabit the earth, is thus to go to the heart of societies and acquire the means to understand them from the inside. It implies resisting the call of a certain fascination with the past, the attractiveness of which lies in the fact that it proposes revisiting a time when things were simpler because they were contained in narrower circles. It is necessary to confront the challenges that technical mutations pose, the increase in mobility, the widening of horizons and the evolution of religious systems and ideologies. Geographers are not satisfied to reconstitute evolutions and to record what works and what constitutes a problem. Their role is to aid people in the creation of forms of social organization and spatial development in accord with present-day possibilities. At the same time, these forms must give a meaning to their lives, provide them with models and allow them to identify with great causes. But the task is not easy!

13.2 Western Uncertainty and Regional Identity

Humanity in search of an identity

Science has demystified nature, has chased away the nymphs of the springs and woods, . . . and heavenly angels are not entered into cosmonauts' computers. This does not prevent the persistence of the great inquiries on origins, life and death.

The land that we spring from, which contains the bodies and the traces of a whole lineage of ancestors in its landscapes, holds a sacred value. The time of the great exodus towards soulless towns is not very old. Links with parents or grandparents who remained '*au pays*', with the stones of the family home or the tombs of the cemetery remain very strong in our societies. For millions of present-day emigrants, for exiles, nostalgia for the land is still alive and the hope to return very keen.

Everywhere modern societies have created contact spaces or economic spaces on the scale of the planet. Mobility becomes a duty, a virtue for the person today who wishes to 'make a career': this contrasts with the past rootedness. All those who count for something – major political or financial leaders, engineers, artists, scholars – belong to a jet-set society. At another level, the charter flights of the Club Mediterranée relocate thousands of middle managers who are delighted to escape from the routine of holidays at their grandmothers' and participate in this way in the globalization of individual destinies.

In fact, everyone participates in a series of contact networks the nodes and links of which are drawn in different spaces; in each of these networks, one plays a role, is known and recognized.

Pierre is of Corsican origin. He is an engineer in the Péchiney aluminium company. His family is based in Lyon. But he spends much of his time between Canada, the United States and Australia, and recently in Dunkirk where the modern factories of his company have been located. In the multinational space of Péchiney he usually speaks English. He is well thought of, his position is good, but in the present recession, it is also fragile. The world price of aluminium can compromise everything. In Lyon, the life of the family district is rather unknown to him. But he plays golf. There he meets industrialists and regional politicians, and this dynamic reality is very obvious to him. The space of Savoy where he owns a chalet is particularly appreciated. The life of the family group is more intense here, and he amazes his children by his prowess in skiing. And Corsica? – a bit remote and yet close. He has inherited a house there in a rather run-down village . . . Why did Péchiney not locate him there? He shrugs his shoulders:

the heart and spirit do not manage the same planet. The bombs of the independence activists doubtless appear derisory to him . . . and mad. Secretly, he understands them a little. Each person's personality is forged according to ceaseless judgements which he is obliged to make between his different roles and his different reference spaces. It is sometimes difficult to be a good father and husband in a work space that is too fragmented!

In the economic field, the life of an individual who seeks to act as a producer or to satisfy his needs as a consumer is linked to a great number of partners. A complex network of hierarchical and interrelated organizations locks them all together; the stability and strength of the ensemble are considerable; many people nevertheless feel uneasy, alone and weak. The place of each person in the ensemble is linked to his effectiveness and his power; but age and competition introduce a perpetual threat of destabilization; even a model employee can be excluded one day.

More or less consciously, everyone exploits relationships between the stable groups he has chosen to belong to or belongs to by right. No risk of exclusion, no exclusive hierarchy, equality in relationships, where solidarity outweighs competition and where emotion normally has its place. It is the close family, or enlarged to an entire clan. It is the larger and chosen fraternal associations, that which is born of a language, a history or shared norms and beliefs. In this same manner, places or spaces that favour this kind of relationship are valued.

Urban regions and identification

The world of immense cities – of giant, blind office towers, of stations, fevered airports, of concrete ribbons of motorways which pierce the empty countrysides – has been forged at the scale of the modern economy. Nobody translates the minuscule place of individuals in this excessive universe better than the cartoonist Sempé.* But in fact, everyone in this cold setting carves niches, full of emotion and strongly exploited.

The basic cell is that of family life. The frightening monotony of the façades of huge cities often hides the extraordinary diversity of tastes and ways of life of the inhabitants. The tyranny of concrete and standardized forms of modern architecture (which sees itself as egalitarian and universal) has practically everywhere replaced the traditional forms of the habitat, adapted to the way of life and structure of family life. The old *quartiers* with narrow streets, with dead ends and blank walls

* A well-known French satirical cartoonist.

of the towns of the Muslim, Hindu or Chinese world, still demonstrate the secrecy surrounding private life, and the discreet place of women.

On a larger scale, the structure of the residential areas is not solely governed by residents' incomes. The American city, with its cultural segregation, Chinatown, Little Italy or black quarter, sometimes seems shocking. In fact, the communities of ethnic or religious origin which favour a certain district are much more diversified than one generally thinks in Europe. This springs from the origin of the settling and the extreme importance of the 'local community' in American civilization. French political philosophy would like all the new arrivals to merge into a common identity. However, we see in all the major agglomerations, and particularly in Paris, the irresistible constitution of Chinatown, North African or African districts, with their shops, their meeting places, pastimes and particular religion. If these social groups are located in notorious squats, from which they are driven out for health or economic reasons, it is a drama. Better to have a tent on a muddy building site than be exiled in a suburban council block, far from one's brothers.

Can the modern town be a welcoming space of collective identification? At the origin of urbanization, the city centre appeared as a sacred space dominated by the imposing masses of the Temple and the Palace. Sacrifices to conciliate the gods; power over humans to assure them of justice and peace, were the conditions of prosperity. These benefits merited a few great collective ceremonies to enthrall the people and convince them of the system's advantages. Our towns still witness certain of these mass demonstrations where the heart of the assembled crowds beats to the unison of shouts and songs. They are called out by trade union or political associations. It is the 'demo' *par excellence*, that fervently expresses the class consciousness of those who feel unjustly treated and struggle for their rights. In this respect, Paris provides traditions, a setting and highly symbolic classic routes. It is said that Haussmann pierced these wide avenues to prevent people from expressing themselves.* But it is thanks to them that they can do it so often and so well!

Other demonstrations gather a more diverse public, perhaps more representative of the whole population. The crowd is capable of filling to suffocation immense stadia, impassioned to the point of hysteria by the imaginary combat in which their town's heroes are engaged. It is the modern form of the combat between the Horaces and the Curiaces,† and passion is expressed no less forcefully. It is on the occasion of

* Haussmann, Prefect of the Département of Seine 1859 to 1870, created boulevards pemitting rapid troop movements and a direct line of fire, as a means of riot control.
† A reference to the battle in the seventh century between the Roman Horace brothers against the three Curiaces, champions of the town of Albe, to determine supremacy.

football (or rugby) matches that today the identification of a town or country is strongest, and this is the case in practically every country.

The great public religious festivals have become rare in Christian countries, but here and there, a few traditional carnivals remain, for the most part taken over by tourism, as at Nice or Rio. In the majority of cases, collective festivals are organized by public authorities to amuse the good folk, to make them forget their misery and give them a semblance of passable artificial coherence. Socialist powers are, or were, very generous in this respect. In communist countries, grandiose parades took the place of banned religious feasts and processions.

But, in fact, urban space has largely lost for its inhabitants the privileged value that was accorded to it by yesterday's citizens. Then, everyone could have the feeling of participating in the virtues of this magical space where the power of gods and great heroes was expressed. It gave to even the most humble people the feeling of being of a superior essence to those of the provinces. Today, a gigantic urbanization appears to many people as an ill of our modern civilization: it is an octopus which strangles us, the tide which gobbles us up, the place of every type of pollution, of all vices and all dangers.

The fantasy of re-ruralization

In contrast, the places where one feels good are close to the fields and woods. In France, the rural roots of town-dwellers are often close. They are rediscovered at holiday times, in the warm atmosphere of grandparents, cousins and near neighbours. This is where one feels at home and where, on retirement, one settles down in the family house. One will be buried in the cemetery alongside the ancestors of all the families of the *pays*. The links between the descendants and the earth remain very alive. But it is no longer the earth as provider. The relaxed and imaginary atmosphere of holidays, when one has time for oneself, where the space one has access to seems larger and freer than in the town, favours a better identification with place. Those who have no heritage seek to find earthly roots by buying old farms to restore or by building new houses in the regional architectural style.

Certain spaces are especially favourable to the blossoming of these new regional identities: seaside and mountains especially, but also the countryside of the interior, rich in lakes and woodlands and the traditional houses or villages which have high aesthetic value. In general terms, this 'imaginary re-ruralization' is a characteristic of developed countries; it affects areas favouring the proliferation of second homes and is accessible to the inhabitants of large agglomerations; they blossom in the countryside where the most distant areas possess a dense network of communications and are served by the public, commercial and health facilities which are considered essential.

The return to nature, the nostalgia for the past, is very good so long as no one is deprived of the benefits of modern civilization. To escape the annoyance of the concentration of mass culture, one seeks to enjoy rarer and more personalized cultural consumption even if it can only be found at the cost of considerable effort and investment. In the name of heritage conservation and the collective passion for old farms or ruined windmills (or, at the coast, for old ports), enormous amounts of public or private capital have been invested. But the fashion goes further: one displays ingeniousness to revive the past more completely. Old ships and old rigging, ploughing implements, old harnesses have been carefully restored to decorate eco-museums or to be paraded in folklore fêtes. Hordes of local historians and ethnologists are active in the work. Costumes, dances and music from the olden days are revived by young enthusiasts who have abandoned rock music. And dialects previously forgotten assume a new vigour. Major public festivities are organized to revive the glorious episodes of regional history.

In an old country like France, where the revolutionary, Jacobin and republican traditions have made an effort to erase the traces of the *ancien régime* and provincial distinctiveness, where French, imposed universally in school, has practically eradicated local dialect, the return to old traditions is more difficult than in neighbouring countries, more recently and less systematically centralized: in central or southern Europe, dialects, feast days and traditional forms of development have been better preserved.

The great fragmentation of rural property, the relative weakness of population density and pressure on the land, the ancient dispersion of the habitat over a high proportion of the country, planning regulations which for a long time were permissive, relatively low construction costs, financial aid for restoration: all of these, on the other hand, have favoured the 'fantasy re-ruralization' in France and permitted thousands of foreigners to participate in it.

In the countries of Eastern Europe, which were submitted in principle to an egalitarian and unifying system, an explosion of local particularities and a strong return to history is evident. But here it is not a matter of a game or an innocent cultural passion, but an explosive and dangerous socio-political evolution.

13.3 The Political Dimension of the Region

The stages of the return of the region

The development of feelings of belonging to a region, in old developed countries like France, appeared initially as a personal matter, a kind of fashion leading to a series of private initiatives in the domain of the environment and culture. But, very quickly, states and public authorities became involved in these localized evolutions.

In a first stage, local elected officials were delighted by this supplement of activities and resources. They turned to the higher levels of administration for subventions, aids and supporting infrastructure (water supply, electricity, telephone, etc.). The state entered into the game. In France, it launched a policy of *pays*. Small urban centres, the capitals of cantons, attempted joint initiatives. The nomenclature of *pays*, devised according to the most varied of logic, blossomed everywhere.

In a second stage, problems arose: contradictory interests appeared. The drama was the success of certain particularly attractive regions: seafronts, mountain valleys and also wooded and lush green zones close to major agglomerations. Prices climbed, for land as well as buildings: a piece of good fortune for those who sold, perhaps too hurriedly, to outsiders; a catastrophe for those who wished to remain in the countryside for they can no longer afford to buy and feel dispossessed of their lands. The agricultural space shrinks, the architectural style becomes bastardized, the landscape is disfigured. Local conflicts are born; one is no longer at home. The newcomers are badly accepted. It is the case of a market town which surges up in a space that was believed to be protected and preserved.

The public authorities are thus called upon to find a solution to conflicts. Planning legislation is demanded to avoid allowing construction 'never mind what, never mind where', and zoning legislation to protect land speculation. Natural Parks protect relatively wild large forested massifs, the high mountains and marshland tracts.

Scarcely before they are in place, the different pieces of legislation are contested and the third phase is entered – that of conflict with the state.

The land is ours!

A first series of conflicts is born from the construction of major public infrastructure. Everyone wants to drive fast but 'no motorways on our land'! Everyone needs electricity supply but 'no pylons nor power stations'. No quarries, no cemeteries, above all no rubbish tips or military camps. One wishes to profit from all the rewards of modern civilization but the inconveniences and risks are good only for other people. The outbreak in the political sphere of ecological ideology makes these conflicts especially relevant.

A second series of conflicts comes from the presence of untolerated newcomers. If it concerns passing guests, seasonal residents, generous clients for businesses or local artisans, all very well. But if they try to compete with the locals, to take away their lands or houses, open businesses, occupy jobs, then mistrust, even open hostility, results.

Live and work in the pays

At this level, imaginary and cultural regionalism, rather jealous and possessive, are not involved. It concerns a fundamental demand which questions the whole of a country's economic system and the distribution in the national space of productive enterprises and employment.

Since the middle of the nineteenth century, the technical upheaval of production, the conflicts and persecutions provoked, in the relatively stable traditional population, widespread uprooting and massive migratory movements on all scales. For a long time, the uprooting was a promise to move towards a better fate, a 'walk to the stars' from the muddy and ruined countryside towards towns full of lights, from old miserable countries towards America or other overseas countries full of promise.

In this light, being rooted seemed an attachment to a worn-out tradition, the refusal of progress. The return to the land and to ancestral virtues seemed a right-wing theme.

Currently, points of view have changed radically. In the name of a seductive but superficial egalitarian ideal, many left-wing politicians have made themselves the champions of regional demands in their most radical form. One ignores, or wishes to ignore, the role played by scale economies and external economies of proximity and the very important place of information and face-to-face relation with economic partners, all of which explains and justifies the driving role of major metropolitan areas in the economy. One demands from the state a deliberate policy of support of traditional enterprises in difficulty, of drip-feeding of subsidized activities in the peripheral regions which suffer from their relative abandonment. It is certain that it is pleasant to live in the Dordogne, in Lozère or in Corsica and that to return to the hurly-burly of the large cities after the holidays is not amusing. But is it reasonable and really effective to wish to demolish everything to solve the problem?

13.4 Independence: Myth or Future for the Regions?

A region is more than a simple administrative region within a state. It is a territory armed with a network of ancient links which are still in existence and, by virtue of its environment, landscapes and population, displays quite strong distinctive features.

A state cannot function unless, beyond these differences, the groups that compose it share a common culture and adhere to shared principles of organization and values. Together, these groups form a nation. The notions of a sovereign people and of democracy can function. The law can effectively be the same for all and the necessary solidarity between all the parties accepted. Any other form of state cannot be maintained except by strong constraint which, in the end, weakens and ruins it.

The conditions in which the division of the world into sovereign states was made and the many regional conflicts which have broken out, internally and at the frontiers, show to what extent the problem is acute at the present time. One might have believed that the socialist ideology here, the parliamentary liberal ideology there, would permit a bonding of the numerous states which issued from the dislocation of colonial empires, that the ideology of human rights, of tolerance, of progress, of science and reason would attenuate traditional cultural and religious contrasts. It has in no way been like that. In many places, people feel unjustly treated and their integrity threatened. They enter into conflict against their state often with the declared objective of independence.

Two series of wrongs, often linked, are invoked: of an ethnic and cultural order on the one hand and economic on the other. The language (or local dialect) is threatened by an imposed national language. Thus an entire traditional culture risks disappearance: that can be felt as an aggression. The religion is threatened by a dominant faith or ideology: the entire system of values and social organization are at stake.

These differences can be accepted in two situations:

(1) Firstly in an imperial situation, where the superiority of the conquering group is so manifest that it is better to be resigned to it so as to escape genocide. If the conquerors display tolerance and avoid interfering in the domain of private life, the traditional chiefs can be submissive or won over. But at the slightest sign of weakness, revolt ensues.

(2) The chances of peaceful coexistence are greatly enhanced in a situation of economic prosperity. If business is going well, criticisms and points of friction diminish. But, in a situation of crisis or manifestly unequal development, nothing works and a strong tendency to introspection and fragmentation rears its head.

What is anticipated of independence apart from certain guarantees and satisfaction of a cultural nature? In some cases, it consists of retaining significant local resources for oneself and managing them for one's own profit. The small petroleum states have this origin, from Brunei to Kuwait, and also some island states, tourist or fiscal paradises, military bases or freeports the success of which is sometimes spectacular, as in the case of Singapore.

But independence is equally claimed by regions that suffer a marginal position, peripheral and underdeveloped within a national space. Since this situation has permitted better conservation of languages or local traditions, all the conditions are united for this type of demand to receive vigorous popular support. In France, this is the case for Corsica and, *a fortiori*, for certain overseas departments and territories. These attitudes seem legitimate, but in fact they can become dramatic.

The movements and mixing of population in the world for several generations have been such that the population located in a particular territory is rarely homogenous and identical in origin. A pseudo-historic right, based on myths or an often biased version of history, is responsible for many merciless massacres like those which have been perpetrated in the Near East for years and at the present time in former Yugoslavia.

The example of Switzerland must be much better studied and reflected on. That would be worthwhile for the older industrialized nations of western Europe and North America, but much more still for Central and Eastern Europe and for the Third World, where the path of development risks being barred by the rise of short-lived regionalisms and tribalisms.

Notes

1 The Development of Regional Studies

1. Jacob, C., *La Description de la terre habitée de Denys d'Alexandre ou la leçon de géographie*, Paris: Michel, 1990, 265 pp., cf. p. 93.
2. Jefferson, T., *Notes on the State of Virginia*, London: Stockdale, 1787.
3. Humboldt, A. de, *Vue des Cordillères et monuments des peuples indigènes de l'Amérique*, Paris: Schoell, 1810.
4. Juillard, E., La région: essai de definition, *Annales de Géographie*, vol. 71, 1962, 483–99.
5. Brunet, R., La composition des modèles dans l'analyse spatiale, *L'Espace Géographique*, vol. 8, no. 4, 1980, 253–65. Brunet, R. and Dolfus, O., *Mondes nouveaux*, Paris: Hachette Reclus, 1990, 551 pp.
6. Gilbert, A., The new regional geography in English and French speaking countries, *Progress in Human Geography*, vol. 12, 1988, 297–316.
7. Claval, P., *Régions, nations, grands espaces: Géographie générale des ensembles territoriaux*, Paris: Genin, 1968, 838 pp.
8. Harvey, D., *Social Justice and the City*, London: Arnold, 1973, 336 pp.
9. Frémont, A., *La Région, espace vécu*, Paris: PUF, 1976, 223 pp.
10. Claval, P., *Eléments de géographie humaine*, Paris: Genin et Litec, 1974, 412 pp.
11. Claval, P., *Principes de géographie sociale*, Paris: Genin et Litec, 1973, 351 pp. Claval, P., *Espace et pouvoir*, Paris: PUF, 1978, 257 pp.
12. Foucault, M., *Surveiller et punir*, Paris: Gallimard, 1975, Trans. *Discipline and Punish: the Birth of the Clinic*, London: Allen Lane, 1977.
13. Dumolard, P., Région et régionalisation: une approche systématique, *L'Espace Géographique*, vol. 4, 1975, 93–111.
14. Dauphiné, A., *Espace, région et système*, Paris: Economica, 1979, 127 pp.
15. Vallega, A., *Regione et Territoria*, Milan: Mursia, 1976, 184 pp. and Vallega, A., *Compendio di geografia regionale*, Milan: Mursia, 1982, 222 pp.

16 Turco, A. (ed.), *Regione et regionalizzazione*, Milan: Franco Angeli, 1984, 303 pp.
17 Auriac, F., *Système économique et espace*, Paris: Economica, 1983.
18 Dicken, P., *Global Shift: The Internationalisation of Economic Activity*, London: Chapman, 2nd edn, 1992, 429 pp.
19 Berque, A., *Vivre l'espace au Japon*, Paris: PUF, 1982, 222 pp.
20 Berque, A., *Le Sauvage et l'artifice: Les Japonais devant la nature*, Paris: Gallimard, 1986, 314 pp.
21 Bonnemaison, J., *Les fondements d'une identité: Territoire, histoire et société dans l'archipel du Vanautu*, vol. 1: *L'arbre et la pirogue*, vol. 2: *Tanna. Les hommes lieux*, Paris: ORSTOM, 1986, 540 pp. and 1987, 618 pp.
22 Pred, A., *Place, Practice and Structure*, Cambridge: Polity Press, 1986, 268 pp.
23 Harvey, D., *The Limits to Capital*, Oxford: Blackwell, 1982.
24 Gregory, D., *Regional Transformation and Industrial Revolution: A Geography of the Yorkshire Woollen Industry*, London: Macmillan, 1982.
25 Gregory, D., Areal differentiation and post-modern human geography. In Gregory, D. and Walford, R. (eds), *Horizons in Human Geography*, London: Macmillan, 1989, pp. 67–96.
26 Giddens, A., *A Contemporary Critique of Historical Materialism*, London: Macmillan, 1981, 294 pp.
27 Giddens, A., *The Constitution of Society*, Cambridge: Polity Press, 1984.
28 Jameson, F., Postmodernism or the cultural logic of late capitalism, *New Left Review*, no. 146, 1984, pp. 53–92.
29 Harvey, D., *The Condition of Postmodernity*, Oxford: Blackwell, 1989, 378 pp.
30 Soja, E., *Postmodern Geographies: The Reassertion of Space in Critical Social Theory*, London: Verso, 1989.
31 Entrikin, J. N., *The Betweenness of Place: Towards a Geography of Modernity*, London: Macmillan, 1991, 196 pp.
32 Cooke, P. (ed.), *Localities: The Changing Face of Urban Britain*, London: Unwin and Hyman, 1989, 320 pp.
33 Johnston, R. J., *A Question of Place: Exploring the Practice of Human Geography*, Oxford: Blackwell, 1991, 280 pp.

2 The Regional Approach

1 Cholley, A., *Guide de l'étudiant en géographie*, Paris: PUF, 1942, 231 pp.
2 Gallois, L., *Régions naturelles et noms de pays*, Paris: Colin, 1908, 356 pp.
3 Claval, P., *Régions, nations, grands espaces: Géographie générale des ensembles territoriaux*, Paris: Genin, 1968, pp. 633–42.
4 Birot, P., *Le Portugal: Etude de géographie régionale*, Paris: Colin, 1949, p. 76.
5 Ibid.
6 Ibid., p. 77.
7 Gallois, 1908.

3 Methods of Regionalization

1 Nystuen, J. D. and Dacey, M. F., A graph theory interpretation of nodal regions, *Papers and Proceedings of the Regional Science Association*, 7, 1961, pp. 29–42.
2 Bunge, W., *Theoretical Geography*, Lund: Gleerup, 1962, 210 pp.
3 Haggett, P., *Locational Analysis in Human Geography*, London: Arnold, 1965.
4 Grigg, D. B., The logic of regional systems, *Annals of the Association of American Geographers*, 55, 1965, pp. 465–91, and Grigg, D. B., Regions, models and classes. In Chorley, R. and Haggett, P. (eds), *Models in Geography*, London: Methuen, 1967, pp. 461–509.
5 Tran Qui Phuoc, *Les régions économiques floues*, Paris: Sirey, 1978, 159 pp.
6 Chisholm, M., *Human Geography: Evolution or Revolution?* Harmondsworth: Penguin, 1975, 207 pp.

7 Regional Consciousness and Identity

1 Elkin, A. P., *Les aborigènes australiens*, Paris: Gallimard, 1967.
2 Bonnemaison, J., Vivre dans l'île, une approche de l'îléité océanienne, *L'Espace Géographique*, vol. 19–20, 1990–1, pp. 119–25.
3 Lévi-Strauss, C., *Anthropologie structurale*, Paris: Plon, 1958.
4 Griaule, M., Mythe de l'organisation du monde chez les Dogons du Soudan, *Psyché*, 6, pp. 443–53.
5 Wheatley, P., *The Pivot of the Four Corners*, Chicago: Aldine, 1971.
6 Rykwert, J., *The Idea of a Town*, London: Faber and Faber, 1976.

9 The Evolution of Forms of Regional Organization: Societies Without a State

1 Malinowski, B., *Les Jardins de Corail*, Paris: Maspéro, 1974.
2 Pourtier, R., *Le Gabon*, vol. 1: *Espace – histoire – société*, Paris: L'Harmattan, 1989.
3 Sautter, G., La région traditionelle en Afrique tropicale, in *Régionalisation et développement*, Paris: CNRS, 1968.
4 Bonnemaison, J., *Les fondements d'une identité: Territoire, histoire et société dans l'archipel de Vanuatu*, vol. 1: *L'Arbre et la Pirogue*, vol. 2: *Tanna. Les hommes lieux*, Paris: ORSTOM, 1986–7.
5 Bonnemaison, J., Vivre dans l'îléité oceanienne, *L'Espace Géographique*, vol. 19–20, 1990–1, pp. 119–25.
6 Malinowski, B., *Les argonautes du Pacifique occidental*, Paris: Gallimard, 1963.

10 The Regional Organization of Traditional Societies

1 Febvre, L., *Philippe II et la Franche-Comté*, Paris: Flammarion, 1970, 538 pp.

2 Sautter, G., La région traditionnelle en Afrique, pp. 65–107, in *Régionalisation et développement*, Paris: CNRS, 1968.
3 Sopher, D., Towards a rediscovery of India: thoughts on some neglected geography. In Mikesell, M. W. (ed.), *Geographers Abroad*, Chicago: Department of Geography Research, paper no. 152, 1973, pp. 110–33.
4 Pelletier, P., Prototypes et archétypes paysagers au Japon. L'exemple du bassin de Nara, *L'Espace géographique*, vol. 16, 1987, pp. 81–93.
5 Polignac, F. de, *La naissance de la cité grecque*, Paris: La Découverte, 1984, 188 pp.
6 Wheatley, P., *The Pivot of the Four Corners*, Chicago: Aldine, 1971, 602 pp.
7 Rykwert, J., *The Idea of the Town: The Anthropology of Urban Form in Rome, Italy and the Ancient World*, London: Faber and Faber, 1976, 242 pp.

Index